Pipefitters and Welders Pocket Manual

Macmillan General Reference
A Simon & Schuster Macmillan Company
1633 Broadway
New York, NY 10019-6785

MACMILLAN is a registered trademark of Macmillan, Inc.

Library of Congress Cataloging-in-Publication Data 96-079620

ISBN: 0-02-034624-7

Manufactured in the United States of America

10 9 8 7 6 5 4 3 2

ACKNOWLEDGEMENTS

The author would like to thank the following companies and individuals for their help in producing *PIPEFITTERS AND WELDERS POCKET MANUAL*.

Jim Atkinson
Atkinson A/C & Refrigeration
6319 Kellum Dr.
Indianapolis, IN 46221

Chemtrol, a Division of NIBCO, Inc.
500 Simpson Ave.
Elkhart, IN 46515

Envirotech Systems
1800 Terminal Rd.
Niles, MI 49120

ITT Fluid Technology Corporation
8200 N. Austin Avenue
Morton Grove, IL 60053

Johnson Controls, Inc.
507 E. Michigan Street
P.O. Box 423
Milwaukee, WI 53201

Kamweld Products Co. Inc.
Norwood, MA 02062

Nibco, Inc.
500 Simpson Avenue
P.O. Box 1167
Elkhart, IN 46515

Praxair, Inc. (formerly Union Carbide Industrial Gases)
39 Old Ridgebury Road
Danbury, CT 06810-5113

Ridge Tool Co.
400 Clark Street
Elyria, OH 44032

Schlumberger Industries
200 Ashford Center North,Suite 200
Atlanta, GA 30338.

Victualic Company of America
P.O. Box 31
Easton, PA 18044-0031

The Viking Corporation
210 N. Industrial Park Road
Hastings, MI 49058

Robinaire Division
SPX Corporation
Montpelier, OH 43543-0193

Watts Regulator
Rte. 114 & Chestnut St.
N. Andover, MA 01845

Special thanks to Bill Unrue, Garry Feliccia, and Shelby Yates of NIBCO, Inc. and Steve Pendergast of Johnson Controls, Inc.

Table of Contents

About the Author

Charles N. McConnell has been a member of the United Association of Journeymen and Apprentices of the Plumbing and Pipefitting Industry of the United States and Canada for over fifty years and has trained many apprentices in the piping trades. He has supervised and installed gas and oil-fired steam and hot water heating systems and air-conditioning installations at Purdue University, in schools, public buildings, and manufacturing facilities. He has supervised and installed process piping systems. He is the author of Home Plumbing Handbook (Macmillan), the three volume Audel Plumbers and Pipe Fitters Library (Macmillan), and Audel's Plumbers Maintenance/Troubleshooting Pocket Manual (Macmillan).

CHAPTER 1

MATH AND METRICS

In the not too distant past, a book such as this would have started with some basic math: addition, subtraction, multiplication, and long division. But with today's apprenticeship classes, on-the-job training, and widespread use of calculators, a chapter devoted to basic math is unnecessary.

Quite often, especially on large construction jobs, job conditions require that offsets in pipe runs be made. Numbers called "constants," taken from trigonometric tables, are used to calculate offsets. These numbers are shown and explained in Chapter 2, Calculating Offsets.

Many times, problems encountered on the job involve the use of fractions. Pipe fitters seldom, if ever, work with fractions other than $1/8$, $1/4$, $3/8$, $1/2$, $5/8$, $3/4$, and $7/8$. Using the decimal equivalents of these fractions is much easier than using the fractions themselves. As a memory aid, Table 1-1 shows the decimal equivalents of fractions from $1/64$ to 1 in.

It is often necessary to find the capacity of a round tank. The more steps involved in solving a problem, the more likely a mistake. The old standard method (with tank measured in inches) is:

FORMULA: $D^2 \times .7854 \times$ Length (L) $\div 231 =$ gallons.
$(D \times D = D^2)$ We'll use our calculator to solve these problems.

$\quad$ D of tank = 12"
$\quad$ L = 60"

Step 1. 12×12 $(D^2) = 144$
Step 2. $144 \times 60 = 8640$
Step 3. $8640 \times .7854 = 6785.856$
Step 4. $6785.856 \div 231 = 29.376$ gallons—(four steps needed)

1

Table 1-1. Decimal Equivalents of Fractions

Inches	Decimal of an Inch	Inches	Decimal of an Inch
1/64	.015625	7/16	.4375
1/32	.03125	29/64	.453125
3/64	.046875	15/32	.46875
1/20	.05	31/64	.484375
1/16	.0625	1/2	.5
1/13	.0769	33/64	.515625
5/64	.078125	17/32	.53125
1/12	.0833	35/64	.546875
1/11	.0909	9/16	.5625
3/32	.09375	37/64	.578125
1/10	.10	19/32	.59375
7/64	.109375	39/64	.609375
1/9	.111	5/8	.625
1/8	.125	41/64	.640625
9/64	.140625	21/32	.65625
1/7	.1429	43/64	.671875
5/32	.15625	11/16	.6875
1/6	.1667	45/64	.703125
11/64	.171875	23/32	.71875
3/16	.1875	47/64	.734375
1/5	.2	3/4	.75
13/64	.203125	49/64	.765625
7/32	.21875	25/32	.78125
15/64	.234375	51/64	.796875
1/4	.25	13/16	.8125
17/64	.265625	53/64	.828125
9/32	.28125	27/32	.84375
19/64	.296875	55/64	.859375
5/16	.3125	7/8	.875
21/64	.328125	57/64	.890625
1/3	.333	29/32	.90625
11/32	.34375	59/64	.921875
23/64	.359375	15/16	.9375
3/8	.375	61/64	.953125
25/64	.390625	31/32	.96875
13/32	.40625	63/64	.984375
27/64	.421875	1	1.0

Now, let's use either of two three-step methods with the same size tank.

Method 1—with tank measured in inches.

$$D = 12"$$
$$L = 60"$$

FORMULA: $D^2 \times L \times .0034 = gallons$

Step 1. $12 \times 12 = 144$

Step 2. $144 \times 60 = 8640$

Step 3. $8640 \times .0034 = 29.376$

Note: Answer is the same, one less step needed, less chance of a mistake.

Method 2—with tank measured in inches and feet.

$$D = 12"$$
$$L = 5'$$

Formula: $D^2 \times L \times .0408 = gallons$

Step 1. $12 \times 12 = 144$

Step 2. $144 \times 5 = 720$

Step 3. $720 \times .0408 = 29.376$

Note: Here again the answer is the same as the others, with one less step needed. As a result, there is less chance of making a mistake.

The above figures can also be used to find the liquid capacity of piping. In the example below use Method 2.

EXAMPLE: How many gallons of foam-water concentrate solution will be needed to fill the following pipe footage?

200 ft. of 6" fire main
500 ft. of 4" fire main
200 ft. of 1" drops to sprinkler heads

Using Method 2 we find:

6 in. $6 \times 6 \times 200 \times .0408 = 293.76$ gals.

4 in. $4 \times 4 \times 500 \times .0408 = 326.40$ gals.

1 in. $1 \times 1 \times 200 \times .0408 = \underline{\quad 8.16}$ gals.

Total gallons needed $\overline{628.32}$ gals

The above formulas can be used in many other calculations performed by pipe fitters.

Table 1-2 is provided as a quick reference for multiplying numbers from 1 to 25.

Working with Metric

The changeover from the English measuring system, which we are all familiar with, to the metric system, which the U.S. government is now adopting, affects all building trades. The principal unit of the metric system is the *meter,* which corresponds roughly to the *yard* as a unit of length. The change to metric in the U.S. has been exceptionally slow. Metric became national policy under the Metric Conversion Act of 1975. Today most packaged products are measured in liters or grams as well as in pints, quarts, gallons, ounces, and pounds. The metric system is used for the legal definition of the yard and pound.

Thinking back to our school days, everyone had a 12 inch ruler. One scale, the one we all used, was marked off in inches. Another scale directly below the inch scale was marked off in metric units such as centimeters and millimeters. However, very few students in the U.S. ever used the metric scale. In many cases, students never had the metric scale or the metric system explained to them. Yet, the metric system, once learned, is simpler and far more accurate than the English system.

Table 1-2. Multiplication Table

No	×1	×2	×3	×4	×5	×6	×7	×8	×9	×10	×11	×12	×13	×14	×15
1	1	2	3	4	5	6	7	8	9	10	11	12	13	14	15
2	2	4	6	8	10	12	14	16	18	20	22	24	26	28	30
3	3	6	9	12	15	18	21	24	27	30	33	36	39	42	45
4	4	8	12	16	20	24	28	32	36	40	44	48	52	56	60
5	5	10	15	20	25	30	35	40	45	50	55	60	65	70	75
6	6	12	18	24	30	36	42	48	54	60	66	72	78	84	90
7	7	14	21	28	35	42	49	56	63	70	77	84	91	98	105
8	8	16	24	32	40	48	56	64	72	80	88	96	104	112	120
9	9	18	27	36	45	54	63	72	81	90	99	108	117	126	135
10	10	20	30	40	50	60	70	80	90	100	110	120	130	140	150
11	11	22	33	44	55	66	77	88	99	110	121	132	143	154	165
12	12	24	36	48	60	72	84	96	108	120	132	144	156	168	180

Table 1-2. Multiplication Table (Cont'd)

No	×1	×2	×3	×4	×5	×6	×7	×8	×9	×10	×11	×12	×13	×14	×15
13	13	26	39	52	65	78	91	104	117	130	143	156	169	182	195
14	14	28	42	56	70	84	98	112	126	140	154	168	182	196	210
15	15	30	45	60	75	90	105	120	135	150	165	180	195	210	225
16	16	32	48	64	80	96	112	128	144	160	176	192	208	224	240
17	17	34	51	68	85	102	119	136	153	170	187	204	221	238	255
18	18	36	54	72	90	108	126	144	162	180	198	216	234	252	270
19	19	38	57	76	95	114	133	152	171	190	209	228	247	266	285
20	20	40	60	80	100	120	140	160	180	200	220	240	260	280	300
21	21	42	63	84	105	126	147	168	189	210	231	252	284	294	315
22	22	44	66	88	110	132	154	176	198	220	242	264	286	308	330
23	23	46	69	92	115	138	161	184	207	230	253	276	299	322	345
24	24	48	72	96	120	144	168	192	216	240	264	288	312	336	360
25	25	50	75	100	125	150	175	200	225	250	275	300	325	350	375

American workmen, especially in the building trades, have always worked with inches, feet, yards, and miles. The transition to the metric system will not be easy; however, the metric system is here and in wide use now, so we might as well make the best of the situation.

Plan measurements formerly shown in feet and inches will, under conversion to metric, be shown in millimeters, centimeters, meters, and kilometers. (They *may* or *may not* be shown in feet and inches as well.) The Construction Metrification Council of the National Institute of Building Sciences in Washington, D.C., states that many federal projects are turning to metric. The following projects are now under way or will be under construction soon.

The Defense Medical Facilities program (a $400 to $500 million project) began the shift to the metric system in 1995.

The National Aeronautics and Space Administration (NASA), the Air Force (all Air Force work since January 1994 has been in metric), the Public Health Service, and the office of the Secretary of Defense are currently working on metric pilot projects with more planned. Estimates vary from $30 million to $60 million in cost.

The General Services Administration had, as of 1993, $1.5 *billion* in planning, design, or construction of projects all using metric. *All work designed by the GSA since January 1994 has been in metric.*

The Department of Energy is building the Super Collider project at a cost over $8.2 *billion*. This project is also being constructed using metric measurements.

The Smithsonian Institute is planning two new facilities with a cost of over $150 million. Both will be built in metric.

A great deal of money is being invested in metric construction projects. The work is there. All construction workers—especially those in the piping trades—will find knowledge of the metric system essential to getting and holding a job. Foremen and job superintendents will be chosen from those with a working knowledge of the metric system as well.

The metric system is based on the "rule of ten." Any basic unit is made larger or smaller by multiplying or dividing it by simple powers of ten. The metric system is a decimal system. In fact, we use it every day when we deal with money. Just as one dollar, divided by ten, equals ten dimes, one meter divided by ten, equals ten decimeters. Under the same principle, one dollar divided by one hundred, equals one hundred cents, as does one meter divided by one hundred, equals one hundred centimeters. So we're already using metric in our money. Construction workers, however, are used to thinking in inches and feet. Changing to metrics will be much simpler if measurements are thought of in terms of tenths and hundredths—or dimes and pennies.

Table 1-3 shows only the metric terms that will affect the work of pipe fitters. This table explains the relationship of metric terms, the symbols for these terms, and the importance of the decimal point in working out metric problems. A misplaced decimal point results in a completely wrong answer.

Table 1-3

1.0 equals a whole number.	Name	Symbol
.1 equals one-tenth of a whole number	deci	d
.01 equals one-hundredth of a whole number	centi	c
.001 equals one one-thousandth of a whole number	milli	m
10.00 equals 10 whole numbers	deka	da
100.00 equals 100 whole numbers	hecto	h
1000.00 equals 1000 whole numbers	kilo	k
Ten millimeters = one centimeter (10mm = 1 cm)		
Ten centimeters = one decimeter (10cm = 1 dm)		
Ten decimeters = one meter (10 dm = 1 m)		
Ten dekameters = one hectometer (10 dam = 1 hm)		
Ten hectometers = one kilometer (10 hm = 1 km)		
One meter = 39.370 inches		
One meter = 3.281 feet		
One meter = 1.093 yard		

Table 1-4 shows the metric to English (or U.S.) equivalents from metric to inches.

Table 1-4. Metric Equivalents

1 Millimeter (mm) (1/1000th of a meter) = .03937 in
10 mm = 1 Centimeter (1/100th of a meter) = .3937 in.
10 cm = 1 Decimeter (1/10th of a meter) = 3.937 in.
10 dm = 1 Meter (1 meter) = 39.370 in.
10 m = 1 Dekameter (10 meters) = 32.8 ft.
10 dm = 1 Hectometer (100 meters) = 328.09 ft.
10 hm = 1 Kilometer (1000 meters) = .62137 mile

Pipe sizes will not change under metrics. All pipe sizes will still be identified as they are now, in inch size, but inside and outside diameters will be shown in both English and metric.

Table 1-5 defines the *meter* and shows the inches equivalents of millimeters.

The *meter* is the base or unit of the system and is defined as one ten-millionth of the distance on the earth's surface from the equator to either pole. Its value in inches should be remembered.
1 meter = 39.37079 inches

Table 1-5. Millimeters into Inches

mm.	inches	mm.	inches	mm.	inches
1/50 =	.00079	26/50 =	.02047	2 =	.07874
2/50 =	.00157	27/50 =	.02126	3 =	.11811
3/50 =	.00236	28/50 =	.02205	4 =	.15748
4/50 =	.00315	29/50 =	.02283	5 =	.19685
5/50 =	.00394	30/50 =	.02362	6 =	.23622
6/50 =	.00472	31/50 =	.02441	7 =	.27559
7/50 =	.00551	32/50 =	.02520	8 =	.31496
8/50 =	.00630	33/50 =	.02598	9 =	.35433
9/50 =	.00709	34/50 =	.02677	10 =	.39370
10/50 =	.00787	35/50 =	.02756	11 =	.43307
11/50 =	.00866	36/50 =	.02835	12 =	.47244
12/50 =	.00945	37/50 =	.02913	13 =	.51181
13/50 =	.01024	38/50 =	.02992	14 =	.55118
14/50 =	.01102	39/50 =	.03071	15 =	.59055
15/50 =	.01181	40/50 =	.03150	16 =	.62992
16/50 =	.01260	41/50 =	.03228	17 =	.66929
17/50 =	.01339	42/50 =	.03307	18 =	.70866
18/50 =	.01417	43/50 =	.03386	19 =	.74803
19/50 =	.01496	44/50 =	.03465	20 =	.78740
20/50 =	.01575	45/50 =	.03543	21 =	.82677
21/50 =	.01654	46/50 =	.03622	22 =	.86614
22/50 =	.01732	47/50 =	.03701	23 =	.90551
23/50 =	.01811	48/50 =	.03780	24 =	.94488
24/50 =	.01890	49/50 =	.03858	25 =	.98425
25/50 =	.01969	1 =	.03937	26 =	1.02362

Table 1-6. Inches into Millimeters

In.	0	1/16	1/8	3/16	1/4	5/16	3/8	7/16
0	0.0	1.6	3.2	4.8	6.4	7.9	9.5	11.1
1	25.4	27.0	28.6	30.2	31.7	33.3	34.9	36.5
2	50.8	52.4	54.0	55.6	57.1	58.7	60.3	61.9
3	76.2	77.8	79.4	81.0	82.5	84.1	85.7	87.3
4	101.6	103.2	104.8	106.4	108.0	109.5	111.1	112.7
5	127.0	128.6	130.2	131.8	133.4	134.9	136.5	138.1
6	152.4	154.0	155.6	157.2	158.8	160.3	161.9	163.5
7	177.8	179.4	181.0	182.6	184.2	185.7	187.3	188.9
8	203.2	204.8	206.4	208.0	209.6	211.1	212.7	214.3
9	228.6	230.2	231.8	233.4	235.0	236.5	238.1	239.7
10	254.0	255.6	257.2	258.8	260.4	261.9	263.5	265.1
11	279.4	281.0	282.6	284.2	285.7	287.3	288.9	290.5
12	304.8	306.4	308.0	309.6	311.1	312.7	314.3	315.9
13	330.2	331.8	333.4	335.0	336.5	338.1	339.7	341.3
14	355.6	357.2	358.8	360.4	361.9	363.5	365.1	366.7
15	381.0	382.6	384.2	385.8	387.3	388.9	390.5	392.1
16	406.4	408.0	409.6	411.2	412.7	414.3	415.9	417.5
17	431.8	433.4	435.0	436.6	438.1	439.7	441.3	442.9
18	457.2	458.8	460.4	462.0	463.5	465.1	466.7	468.3
19	482.6	484.2	485.8	487.4	488.9	490.5	492.1	493.7
20	508.0	509.6	511.2	512.8	514.3	515.9	517.5	519.1
21	533.4	535.0	536.6	538.2	539.7	541.3	542.9	544.5
22	558.8	560.4	562.0	563.6	565.1	566.7	568.3	569.9
23	584.2	585.8	587.4	589.0	590.5	592.1	593.7	595.3

Table 1-6. Inches into Millimeters (Cont'd)

In.	1/2	9/16	5/8	11/16	3/4	13/16	7/8	15/16
0	12.7	14.3	15.9	17.5	19.1	20.6	22.2	23.8
1	38.1	39.7	41.3	42.9	44.4	46.0	47.6	49.2
2	63.5	65.1	66.7	68.3	69.8	71.4	73.0	74.6
3	88.9	90.5	92.1	93.7	95.2	96.8	98.4	100.0
4	114.3	115.9	117.5	119.1	120.7	122.2	123.8	125.4
5	139.7	141.3	142.9	144.5	146.1	147.6	149.2	150.8
6	165.1	166.7	168.3	169.9	171.5	173.0	174.6	176.2
7	190.5	192.1	193.7	195.3	196.9	198.4	200.0	201.6
8	215.9	217.5	219.1	220.7	222.3	223.8	225.4	227.0
9	241.3	242.9	244.5	246.1	247.7	249.2	250.8	252.4
10	266.7	268.3	269.9	271.5	273.1	274.6	276.2	277.8
11	292.1	293.7	295.3	296.9	298.4	300.0	301.6	303.2
12	317.5	319.1	320.7	322.3	323.8	325.4	327.0	328.6
13	342.9	344.5	346.1	347.7	349.2	350.8	352.4	354.0
14	368.3	369.9	371.5	373.1	374.6	376.2	377.8	379.4
15	393.7	395.3	396.9	398.5	400.0	401.6	403.2	404.8
16	419.1	420.7	422.3	423.9	425.4	427.0	428.6	430.2
17	444.5	446.1	447.7	449.3	450.8	452.4	454.0	455.6
18	469.9	471.5	473.1	474.7	476.2	477.8	479.4	481.0
19	495.3	496.9	498.5	500.1	501.6	503.2	504.8	506.4
20	520.7	522.3	523.9	525.5	527.0	528.6	530.2	531.8
21	546.1	547.7	549.3	550.9	552.4	554.0	555.6	557.2
22	571.5	573.1	574.7	576.3	577.8	579.4	581.0	582.6
23	596.9	598.5	600.1	601.7	603.2	604.8	606.4	608.0

CHAPTER 2

CALCULATING OFFSETS

Table 2-1. Multipliers for Calculating Simple Offsets

To Find Side	When Known Side Is	Multiply Side	Using 60° Elbows by	Using 45° Elbows by	Using 22½° Elbows by	Using 11¼° Elbows by
T	S	S	1.155	1.414	2.613	5.125
S	T	T	.866	.707	.383	.195
R	S	S	.577	1.000	2.414	5.027
S	R	R	1.732	1.000	.414	.198
T	R	R	2.000	1.414	1.082	1.019
R	T	T	.500	.707	.924	.980

Most of the common offsets used in piping runs are made with 45°, 60°, and 11¼° fittings, in that order. The constants shown in Table 2-1 are used to calculate these offsets. Examples are shown below. Fig 2-1 shows a simple offset, using 45° elbows. To find the *travel,* the center-to-center measurement between two angled fittings when the *set* is known, the multiplier for the angle fitting is used.

Example 1: What is the length of side T for a 45° offset if side S is 19½ in.?

The side T is found in the first column of Table 2-1. The known side, S, is shown in the second column of Table 2-1. The constant, or multiplier for 45° offsets, 1.414, is shown in the fifth column.

Answer: T = S × 1.141

T = 19½ × 1.414 = 27.573 in. (round off to 27½ in.)

Example 2: What is the length of side S if the travel is 19 in.?

The known side, T, is shown in column 2 of the Table.

$$S = T \times .707 = 13.433 \text{ in. (round off to } 13\frac{1}{2} \text{ in.)}$$

Example 3: What is the length of side R if the set is 10 in.?

$$R = S \times 1.000 = 10 \text{ in.}$$

Example 4 (Using 60° fittings): What is the length of side T if side S is 19¹/₂ in.?

The known side, S, is shown in column 2 of the Table.

$$T = 19\frac{1}{2} \times 1.155 = 22.522 \text{ in. (round off to } 22\frac{1}{2} \text{ in.)}$$

Example 5: What is the length of side S if side T is 8 in.?

$$S = 8 \times .866 = 6.92 \text{ in. (round off to 7 in.)}$$

Example 6: What is the length of side R if side S is 8¹/₂ in.?

$$R = 8\frac{1}{2} \times .577 = 4.90 \text{ in. (round off to 5 in.)}$$

To find *set* when run is known, *travel* when run is known, and *run* when travel is known, use the same procedure shown in examples above.

The calculated measurements using the constants shown are *center-to-center measurements.* Deduct the end-to-center measurements of the fittings (X in inset of Fig. 2-1) to get the end-to-end measurement of the cut pipe.

The *set, run,* and *travel* for an 11¹/₄°, and a 22¹/₂° offset can also be calculated from the constants shown in Table 2-1.

Example 1. What is the *set* (S) for a 22¹/₂° offset if the *travel* (T) is 15 in.?

Answer: Multiply side T (15) × .383 = 5.74. Round to 5 ³/₄ in.
          The set is 5 ³/₄ in.

Example 2: What is the *run* (R) of a 22¹/₂° offset if the *set* (S) is 8 in.?

Answer: Multiply side S (8) × 2.414 = 19.31. Round to 19¹/4. The run is 19¹/4 in.

Example 3: What is the *travel* of a 22¹/2° offset if the *run* is 10 in.?

Answer: Multiply side R × 1.082 = 10.82. Round to 10³/4. The travel is 10³/4 in.

Table 2-2. Relationships of Set and Travel in 45° Offsets

Set	Travel	Set	Travel	Set	Travel
2	2.828	9	12.726	16	22.624
¹/4	3.181	¹/4	13.079	¹/4	22.977
¹/2	3.531	¹/2	13.433	¹/2	23.331
³/4	3.888	³/4	13.786	³/4	23.684
3	4.242	10	14.140	17	24.038
¹/4	4.575	¹/4	14.493	¹/4	24.391
¹/2	4.949	¹/2	14.847	¹/2	24.745
³/4	5.302	³/4	15.200	³/4	25.098
4	5.656	11	15.554	18	25.452
¹/4	6.009	¹/4	15.907	¹/4	25.705
¹/2	6.363	¹/2	16.261	¹/2	26.059
³/4	6.716	³/4	16.614	³/4	26.412
5	7.070	12	16.968	19	26.866
¹/4	7.423	¹/4	17.321	¹/4	27.219
¹/2	7.777	¹/2	17.675	¹/2	27.573
³/4	8.130	³/4	18.028	³/4	27.926
6	8.484	13	18.382	20	28.280
¹/4	8.837	¹/4	18.735	¹/4	28.635
¹/2	9.191	¹/2	19.089	¹/2	28.987
³/4	9.544	³/4	19.442	³/4	29.340
7	9.898	14	19.796	21	29.694
¹/4	10.251	¹/4	20.149	¹/4	30.047
¹/2	10.605	¹/2	20.503	¹/2	30.401
³/4	10.958	³/4	20.856	³/4	30.754
8	11.312	15	21.210	22	31.108
¹/4	11.665	¹/4	21.563	¹/4	31.461
¹/2	12.019	¹/2	21.917	¹/2	31.815
³/4	12.382	³/4	22.270	³/4	32.168

continues

Table 2-2. Relationships of Set and Travel in 45° Offsets (Cont'd)

Set	Travel	Set	Travel	Set	Travel
23	32.522	32	45.248	41	57.974
1/4	32.875	1/4	45.601	1/4	58.327
1/2	33.229	1/2	45.955	1/2	58.681
3/4	33.582	3/4	46.308	3/4	59.034
24	33.936	33	46.662	42	59.388
1/4	34.279	1/4	47.015	1/4	59.741
1/2	34.643	1/2	47.369	1/2	60.095
3/4	34.996	3/4	47.722	3/4	60.448
25	35.350	34	48.076	43	60.802
1/4	35.703	1/4	48.429	1/4	61.155
1/2	36.057	1/2	48.783	1/2	61.509
3/4	36.410	3/4	49.136	3/4	61.862
26	36.764	35	49.490	44	62.216
1/4	37.117	1/4	49.843	1/4	62.569
1/2	37.471	1/2	50.197	1/2	62.923
3/4	37.824	3/4	50.550	3/4	63.276
27	38.178	36	50.904	45	63.630
1/4	38.531	1/4	51.257	1/4	63.983
1/2	38.885	1/2	51.611	1/2	64.337
3/4	39.238	3/4	51.964	3/4	64.690
28	39.592	37	52.318	46	65.044
1/4	39.945	1/4	52.671	1/4	65.397
1/2	40.299	1/2	53.025	1/2	65.751
3/4	40.652	3/4	53.378	3/4	66.104
29	41.006	38	53.732	47	66.458
1/4	41.359	1/4	54.085	1/4	66.811
1/2	41.713	1/2	54.439	1/2	67.165
3/4	42.066	3/4	54.792	3/4	67.518
30	42.420	39	55.146	48	67.872
1/4	42.773	1/4	55.499	1/4	68.225
1/2	43.127	1/2	55.853	1/2	68.579
3/4	43.480	3/4	56.206	3/4	68.932
31	43.834	40	56.560	49	69.286
1/4	44.187	1/4	56.913	1/4	69.639
1/2	44.541	1/2	57.267	1/2	69.993
3/4	44.894	3/4	57.620	3/4	70.346

Table 2-2. Relationships of Set and Travel in 45° Offsets (Cont'd)

Set	Travel	Set	Travel	Set	Travel
50	70.700	59	83.426	68	96.152
1/4	71.053	1/4	83.779	1/4	96.505
1/2	71.407	1/2	84.133	1/2	96.859
3/4	71.760	3/4	84.486	3/4	97.212
51	72.114	60	84.840	69	97.566
1/4	72.467	1/4	85.193	1/4	97.919
1/2	72.821	1/2	85.547	1/2	98.273
3/4	73.174	3/4	85.900	3/4	98.626
52	73.528	61	86.254	70	98.980
1/4	73.881	1/4	86.607	1/4	99.333
1/2	74.235	1/2	86.961	1/2	99.687
3/4	74.588	3/4	87.314	3/4	100.040
53	74.942	62	87.668	71	100.394
1/4	75.295	1/4	88.021	1/4	100.747
1/2	75.649	1/2	88.375	1/2	101.101
3/4	76.002	3/4	88.728	3/4	101.454
54	76.356	63	89.082	72	101.808
1/4	76.709	1/4	89.435	1/4	102.165
1/2	77.063	1/2	89.789	1/2	102.515
3/4	77.416	3/4	90.142	3/4	102.868
55	77.770	64	90.496	73	103.222
1/4	78.123	1/4	90.849	74	104.636
1/2	78.477	1/2	91.203	75	106.050
3/4	78.830	3/4	91.556	76	107.464
56	79.184	65	91.910	77	108.878
1/4	79.537	1/4	92.263	78	110.292
1/2	79.891	1/2	92.617	79	111.706
3/4	80.244	3/4	92.970	80	113.120
57	80.598	66	93.324	81	114.534
1/4	80.951	1/4	93.677	82	115.948
1/2	81.305	1/2	94.031	83	117.362
3/4	81.658	3/4	94.384	84	118.776
58	82.012	67	94.738	85	120.190
1/4	82.365	1/4	95.091	86	121.604
1/2	82.719	1/2	95.445	87	123.018
3/4	83.072	3/4	95.798	88	124.432

continues

Table 2-2. Relationships of Set and Travel in 45° Offsets (Cont'd)

Set	Travel	Set	Travel	Set	Travel
89	125.846	100	141.400	111	156.954
90	127.260	101	142.814	112	158.368
91	128.674	102	144.228	113	159.782
92	130.088	103	145.672	114	161.196
93	131.502	104	147.056	115	162.610
94	132.916	105	148.470	116	164.024
95	134.330	106	149.884	117	165.438
96	135.744	107	151.298	118	166.852
97	137.158	108	152.712	119	168.266
98	138.572	109	154.126	120	169.680
99	139.986	110	155.540		

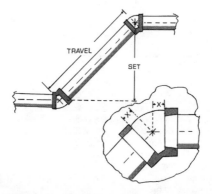

Fig. 2-1. A 45° simple piping offset.

The X in the inset shows the take-off (allowance) from the end of the thread in the fitting to the center of the fitting.

Two-Pipe 45° Equal-Spread Offset

A two-pipe equal-spread offset is shown in Fig. 2-2.

The steps for working out the dimensions for a two-pipe 45° equal-spread offset are:

FORMULA: A = spread
            S = set
            T = S × 1.414
            R = S × 1.000
            F = A × .4142
            D = T

EXAMPLE: Find run (R) and travel (T) when the spread is 6 in. and the set is 8 in.
            A = 6 in.
            S = 8 in.
            T = 8 × 1.414 = 11.31 in. (round to 11 1/4 in.)
            R = 8 × 1.000 = 8 in.
            F = 6 × .4142 = 2.48 in. (round to 2 1/2 in.)
            D and T are the same length.
            The run is 8 in. and the travel is 11 1/4 in.

Note: When the spreads of the pipes are the same, pipe D is always .41 times the spread longer than the other pipe.

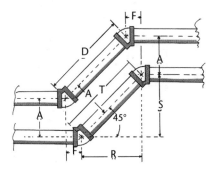

Fig. 2-2. A two-pipe 45° equal-spread offset.

A Two-Pipe 45° Equal Spread Offset at a Corner

A two-pipe equal-spread offset at a corner is shown in Fig. 2-3. This kind of problem is solved by following the steps shown below. The spread, (in the example shown, 6 in.) is the key to solving the problem.

EXAMPLE: A = 9 × .4142 = 3.72 in. (round to 3³/₄ in.)
                      B = 18 × 1.414 = 25.45 in. (round to 25¹/₂ in.)
                      C = A + A + B = 3³/₄ + 3³/₄ + 25¹/₂ = 33 in.

Pipe D is always .41 times the spread longer than the other pipe when the spreads of the pipes are the same.

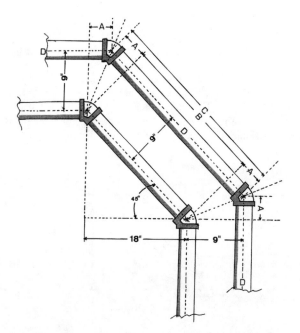

Fig. 2-3. A two-pipe 45° equal-spread offset at a corner.

Finding Travel and Run for a Two-Pipe 22½° Equal-Spread Offset

A 22½ ° equal-spread offset is shown in Fig. 2-4. The formula for solving a problem of this kind is shown below.

EXAMPLE: Find the travel and run for a two-pipe 22½° equal-spread offset if the spread is 8 inches.

FORMULA: A = spread = 8 in.

S = set = 10 in.

T = travel

R = run

$F = A \times .1989$

$T = S \times 2.613 = 10 \times 2.613 = 26.13$ in. (round to 26 ⅛ in.)

$R = S \times 2.414 = 10 \times 2.414 = 24.14$ in. (round to 24 ⅛ in.)

$F = A \times .1989 = 8 \times .1989 = 1.59$ in. (round to 1½ in.)

The travel is 26⅛ in.

The run is 24⅛ in.

D and T are the same length in this type problem.

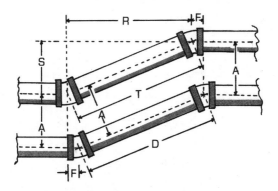

Fig. 2-4. A two-pipe 22½° equal-spread offset.

Finding the Starting Point of a 45° Offset Around a Pilaster or Column

Fig. 2-5 shows a 45° offset around a pilaster or column. The formula below shows how to find the starting point of the offset.

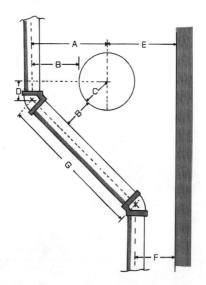

Fig. 2-5. Finding starting point for a 45° offset around a pilaster.

FORMULA: A = distance from center of pilaster to center line of pipe.
B = distance from side of pilaster to center of pipe.
C = one-half of the diameter of the pilaster.
D = distance from center line of pilaster to starting point of offset.

E = distance from center line of pilaster to wall.
F = distance from center of pipe to wall.
A = B + C
D = A × .4142

EXAMPLE: Find D if C is 14 in. and B is 10 in.
A = B + C = 10 + 14 = 24 in.
D = A × .4142 = 9.94 in. (round to 10 in.)
Therefore the center of the 45° elbow is 10 in.
from the center of the pilaster.
E = 10 in.
F = 8 in.
G = A + E − F × 1.414 = 24 + 10 − 8 × 1.414 =
36.76 in. (36 3/4 in.)

Three-pipe 45° Equal-Spread Offset Around a Pilaster or Obstruction

The first step in working out a problem of the type shown in Fig. 2-6 is to locate the starting point using A and B in the formula below. With the starting point located, the lengths of C, D, and E are easily figured.

Dimensions of piping layout shown in Fig. 2-6:

A = Diameter of pilaster + 12 + 8 = 40 in.
(20 + 12 + 8 = 40)
B = Radius of pilaster + 8 × .4142 = 7.4556
(10 + 8 × .4142) = 7.4556
C = A − 8 × 1.414 = 45.248 (40 − 8 × 1.414 = 45.248)
D = A + 8 − 16 × 1.414 = 45.248 (40 + 8 − 16 × 1.414)
E = A + 8 + 8 − 24 × 1.414 = 45.248
(40 + 8 + 8 − 24 × 1.414)
F = 8 × .4142 = 3.316 in. (3 5/16 in.)
G = 8 + 8 × .4142 = 6.627 in. (6 5/8 in.)

Note: In a problem of this type, pipes C, D, and E are the same length.

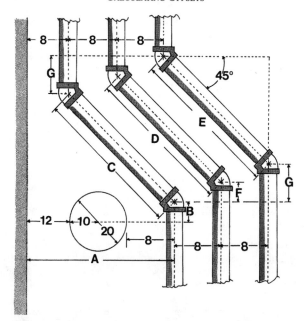

Fig. 2-6. A three-pipe 45° equal-spread offset around a pilaster.

Finding the Starting Point for a 45° Offset at a Corner

Fig. 2-7 shows an offset in a wall. A pipe run following the wall must be offset to follow the wall. The starting point for the offset can be found using the following formula:

A = distance from wall to starting point of offset.

B = distance from corner to center line of run of pipe.

C = distance from corner to center of pipe.

D = distance from wall to center of pipe after offset. (14 in.)

E = distance from wall before offset. (8 in.)

A = B + (C × 1.414)

EXAMPLE: Find distance A if B is 12 in. and C is 8 in.

A = B + (C × 1.414) = A = 12 + 11.31 = A = 23.31 in. (23¹/₄ in.)

Therefore the center of the 45° elbow is 23¹/₄ in. from the wall.

F = D – E × 1.414 = F = 14 – 8 × 1.414 = F = 6 × 1.414 = 8.48 in.

Rounding the 8.48 to 8¹/₂ in., F = 8¹/₂ in.

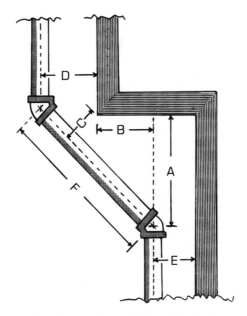

Fig. 2-7. Finding the starting point for a 45° offset at a corner.

45° Unequal-Spread Offsets

A 45° unequal-spread offset is shown in Fig. 2-8. The formulas for working out a problem of this kind are shown below.

FORMULAS: A = spread No. 1
B = spread No. 2
C = spread No. 3
D = spread No. 4
$E = A \times 1.414$
$F = E - C$
$G = F \times 1.414$
$H = A - G$
$J = B \times 1.414$
$K = D - J$
$L = K \times 1.414$
$M = L + B + H$

EXAMPLE: Find the lengths of H and M for a 45° unequal-spread offset when:
A = 12 in.
B = 8 in.
C = 12 in.
D = 15 in.
$E = 12 \times 1.414 = 16.96$ in. (round to 17 in.)
$F = 17 - 12 = 5.00$ in.
$G = 5 \times 1.414 = 7.07$ in. (round to 7 in.)
$H = 12 - 7 = 5.00$ in.
$J = 8.00 \times 1.414 = 11.31$ in. (round to $11^1/4$ in.)
$K = 15 - 11^1/4 = 3\,^3/4$ in. (3.75)
$L = 3.75 \times 1.414 = 5.30$ in. (round to $5^1/4$ in.)
$M = 5^1/4 + 8 + 7.00 = 20\,^1/4$ in.
Answer:
H = 5 in.
M = $20^1/4$ in.

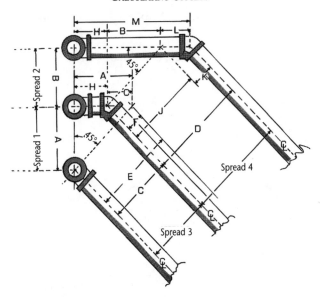

Fig. 2-8. A three-pipe 45° unequal-spread offset.

Finding Travel and Run for a Two-Pipe 60° Equal-Spread Offset

A two-pipe 60° equal-spread offset is shown in Fig. 2-9. Using the formula shown below, find the travel (T) and run (R) for a 60° offset if the set is 10 in. and the spread is 8 in.

FORMULA: A = spread
             S = set
EXAMPLE: T = S × 1.155 = 10 × 1.55 = 11.55 in. (round to 11½ in.)
             R = S × .5773 = 10 × .5773 = 5.77 in. (round to 5¾ in.)

$F = A \times .5773 = 8 \times .5773 = 4.61$ in. (round to $5\frac{7}{8}$ in.)

Therefore the travel is $11\frac{1}{2}$ in., the run is $5\frac{3}{4}$ in.

D and T are the same length in this type problem.

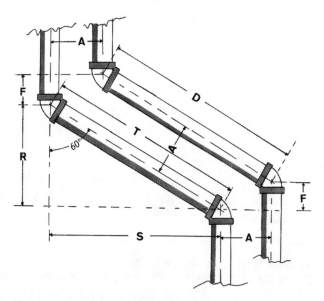

Fig. 2-9. A two-pipe 60° equal-spread offset.

FORMULA: A = spread
S = set
$T = S \times 1.155$
$R = S \times .5773$
$F = A \times .5773$
D and T are same length.

Finding Travel and Run for a 45° Rolling Offset

A 45° rolling offset is shown in Fig. 2-10. The formula for figuring a rolling offset is shown below.

FORMULA: $A = \sqrt{roll^2 + set^2}$
Travel = A × cosecant of angle of fitting.
Run = A × cotangent of angle of fitting.
(Refer to Trigonometry Table)

EXAMPLE: The roll of a 45° offset is 9 in. and the set is 12 in. What is the length of travel and run?

$A = \sqrt{roll^2 + set^2}$

$A = \sqrt{81 + 144} = \sqrt{225} = 15$ in.

Travel = A × cosecant of angle of fitting
Travel = 15 × 1.414 = 21.21 in. (21 1/4 in.)
Run = A × cotangent of angle of fitting
Run = 15 × 1.000 = 15 in. center-to-center

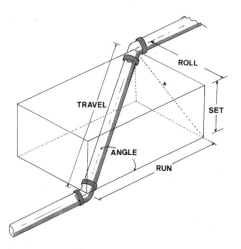

Fig. 2-10. Finding travel and run for a 45° rolling offset.

CHAPTER 3

PIPE WELDING WITH OXYACETYLENE AND ARC

OXYACETYLENE WELDING

This section is designed to aid the novice who wants to learn the techniques needed to produce good oxyacetylene welds. Many highly skilled welders have learned this technique by picking up a torch and practicing on scrap pipe until they became proficient. The only way to learn welding is through the "hands on" method—it cannot be learned from a book. However, if a person has a basic understanding of the subject (knowing how to hold the torch and welding rod at the correct angle for various positions as the weld is being made), the job can be easy. A list of very important precautions to follow when using oxyacetylene equipment will be found at the end of this section.

Setting Up the Equipment

Oxygen and acetylene tanks should either be mounted and secured to a cart or securely fastened to a bench, wall, or post. These cylinders should never be stored or used in any position other than an upright one. Before mounting the gauges on the tanks, open the valves $1/4$ turn and then quickly close them (Fig. 3-1). This is called "cracking" the valves and will blow out any foreign matter that might otherwise get into the gauges.

Connect the oxygen regulator (Fig. 3-2) to the oxygen tank and the acetylene regulator (Fig. 3-3) to the acetylene tank. Two different kinds of acetylene cylinder connections are used in the

U.S.: the CGA 510 connection has left hand threads, internal on the cylinder outlet; the CGA 300 connection has right hand threads, external on the cylinder outlet. Tighten both connections firmly with a wrench.

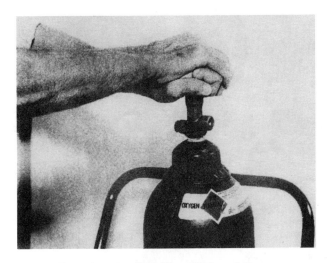

Fig. 3-1. The safe way to "crack" a cylinder valve.

To close the valve, rotate the pressure-adjusting screw on each regulator to the left, counterclockwise, until it turns freely. The regulator valves must be closed before pressure is applied to the regulators. Stand to the side but where you can see the regulator, and open the valves *slowly*. The oxygen valve has a wheel handle and should be turned to full open by turning it to the left, counterclockwise. The acetylene valve also must be turned to the left or counterclockwise. The acetylene valve may have a wheel handle or may require a wrench to open it. In either case, it should be opened only 1/4 turn, and if a wrench is used, the wrench should

LOW PRESSURE
GAUGE

HIGH PRESSURE
GAUGE

PRESSURE
ADJUSTING
SCREW

HOSE
CONNECTION

HIGH PRESSURE
INLET

Courtesy Veriflo Corp.

Fig. 3-2. A two-stage oxygen regulator.

be left on the valve as shown in Fig. 3-4 in order for the valve to be turned off quickly in case of an emergency. When the valves are opened, full tank pressure will be shown on the high pressure gauges.

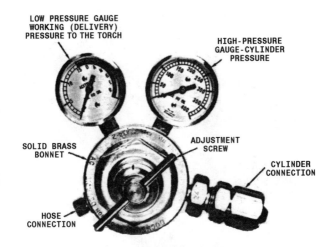

LOW PRESSURE GAUGE
WORKING (DELIVERY)
PRESSURE TO THE TORCH

HIGH-PRESSURE
GAUGE-CYLINDER
PRESSURE

SOLID BRASS
BONNET

ADJUSTMENT
SCREW

CYLINDER
CONNECTION

HOSE
CONNECTION

Courtesy Dockson Corp.

Fig. 3-3. A two-stage acetylene regulator.

Oxygen and acetylene hoses are made specifically for gas welding and cutting purposes. The oxygen hose is green while the acetylene hose is red. The oxygen hose has right hand coupling nuts for connecting to the regulator and welding barrel. Acetylene coupling nuts have left hand threads. When connecting the oxygen and acetylene hoses, the coupling nuts should thread on easily with only a slight tightening with a wrench needed. Oxygen and acetylene connections are made with a ground joint; *do not use thread lubricants, oil, or grease on these connections.*

A typical welding torch (A) and a typical cutting head (B) are shown in Fig. 3-5. Both the welding torch and the cutting head can be mounted on the same barrel. The barrel contains the needle valves controlling both the oxygen and acetylene supply to the welding torch and the cutting torch. When the cutting head is used,

Fig. 3-4. Always leave a wrench on an acetylene cylinder valve when tank is in use.

the needle valve on the head controls the oxygen supply to the cutting (or large) orifice in the head. The small orifices in the cutting head are used to bring the metal up to heat for cutting. With all connections made, turn off the needle valves on the welding torch barrel by turning the valve knobs clockwise. Then open the oxygen regulator by turning the screw on the regulator clockwise and set the pressure on the low pressure gauge to approximately 25 psig (lbs. sq. in. gauge). Then, open the acetylene regulator by turning the screw on the regulator clockwise until a setting of 7 to 10 psig

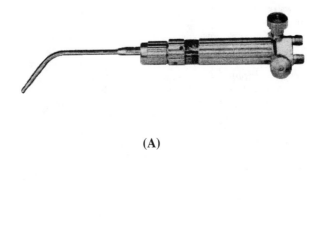

(A)

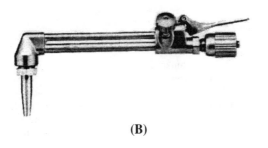

(B)

Fig. 3-5. A typical welding torch (A) and a cutting head (B).

shows on the low pressure gauge. You will notice that the low pressure acetylene regulator is painted red above 15 psig. Because acetylene becomes unstable at a working pressure above 15 psig, *acetylene must never be used at a pressure above 15 psig.*

Test all connections for leaks by turning off the oxygen and acetylene supplies at the regulators. Turn the screws on the regulators counterclockwise, loosen them, and check that the needle valve knobs on the torch barrel are turned off. This leaves pressure on the hoses between the regulator and the welding torch barrel. If there is a leak at a hose connection, the pressure on the low pressure gauges will drop.

CAUTION: If it should ever be necessary to retighten a coupling nut after the outfit has been set up, close the cylinder valve before tightening the nut.

Eye Protection

Welding goggles or a welding hood with shaded glass must be worn when acetylene cutting or welding is being done. Shaded glass in the color range of 4 to 6 is generally used.

Lighting the Torch for Welding

With the welding tip mounted on the torch barrel, open the acetylene valve on the torch about 1/4 turn and *immediately* light the flame with a friction lighter. *Never use a match!* Reduce the acetylene flow by throttling the torch acetylene valve until the flame starts to produce black smoke. Then increase acetylene flow until the smoke disappears. Open the oxygen valve slowly until a neutral flame appears. A neutral flame is shown in Fig. 3-6. To adjust the volume of the flame, alternately increase the acetylene and oxygen. It is important to use the correct size welding tip for the job in hand. If the tip is too large and must be throttled back to produce the flame volume needed, the torch will "pop" and "backfire." If the flame burns away from the tip, throttle back on the oxygen, then the acetylene, maintaining a neutral flame until the flame returns to the tip. If the flame goes out and burns back within the torch, creating a shrill whistling noise, *turn the torch off immediately.* This is called a "flashback" and indicates that something is wrong, either with the torch or with operation of the torch. Allow the torch to cool before attempting to relight it.

PRACTICING ON PIPE

The first step in making the weld is to cut and bevel two pieces of pipe, each piece 8" or 10" long. A soapstone crayon can be used to mark the cut. The working pressures on the oxygen and acetylene gauges should be set as described earlier. Mount the cutting head on the torch barrel. The cutting torch (Fig. 3-5) should be ignited and adjusted for a soft neutral flame using the needle valve knobs on the *barrel*. (A soft flame is one that has a long blue inner cone.) The needle valve on the cutting *head* should be opened by turning the knob counter-clockwise. The torch should be held at an angle as shown in Fig. 3-7 in order to bevel the pipe. When the metal is red hot, depress the cutting lever (applying oxygen at secondary gauge pressure) and move the cutting torch around the edge, rolling the pipe as the cut is made. Use a chipping hammer (Fig. 3-8) to remove any slag at the cut edge. Lay the pieces in an angle or channel iron as shown in Fig. 3-9, leaving a $1/16$" gap between the pieces and make four tack welds 90° apart.

Making the Tack Welds

Mount the welding tip on the barrel. A #3 (drill size 53) or #4 (drill size 49) tip should be used for welding on 4" or 6" pipe along with a $1/8$" mild steel welding rod. Light the torch and adjust for a soft neutral flame. The flame volume will depend on the size tip being used. Generally the flame should be as large as the welding tip will permit without the soft blue tip of a neutral flame pulling away from the end of the welding tip. A tack weld is a single small weld made at the bottom of the beveled edges of the pieces to be welded. The tack weld is made by playing the torch flame on both pieces, heating the metal until it is white hot, then touching the welding rod to one side. A pool of melted metal will form and flow across to the other piece. Withdraw the torch and turn the

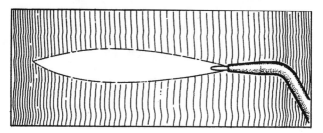

A neutral welding flame.

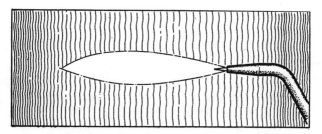

An oxidizing welding flame.

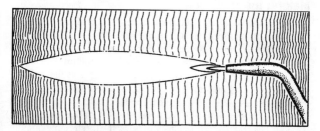

A carburizing welding flame.

Fig. 3-6. Three basic welding flames.

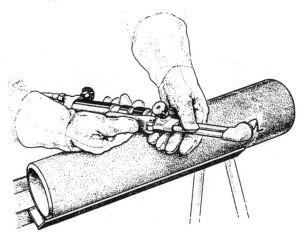

Fig. 3-7. Correct position for making a beveled cut.

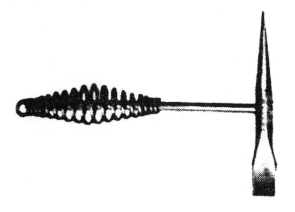

Fig. 3-8. A chipping hammer.

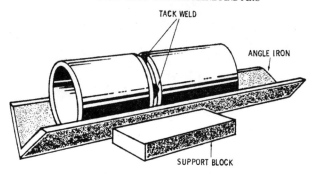

Fig. 3-9. Using angle iron to align pipe for welding.

pipe 180° and make a tack there. Make two more tacks midway between the first two. There are now four tacks, 90° apart, and the pipe is ready for the finish weld.

Forehand and Backhand Technique

Both the forehand and backhand techniques are shown in Fig. 3-10. When the forehand method is used, the welding rod moves ahead of the torch tip. When the backhand method is used, the torch moves ahead of the welding rod. In pipe welding when the pipe cannot be rotated, the welder must be able to switch easily from forehand to backhand. The forehand technique is used to weld from bottom to top of the pipe. The backhand technique is used to weld from top to bottom of the pipe. With either technique, the flame is directed almost, but not quite, to the center line of the pipe. This angle may have to be changed slightly to control the molten "puddle" in the horizontal and overhead positions. The rod must rub in the bottom of the puddle, touching each side of the gap. As rod is added to the puddle, the flame and rod should move in opposite directions with the flame directed at one side and the rod touching the other side in a "whipping" or U-shaped motion. The flame must be directed at the sides of the gap just long enough to ensure that the side metal is melted to allow fusion and complete penetration.

Making a Rolling Weld

Light the torch and set it for a neutral flame. Then start about 2" down from the top (at the 1:30 position in Fig. 3-11) and, using a U motion with the torch tip, heat the area white hot. When the metal starts to melt, insert the welding rod and form a pool of molten metal. Add welding rod, keeping the end of the rod in the pool, and build the pool up to the top of the pipe. As the pipe is rolled counterclockwise (down), the tip should stay at the 1:30 position (Fig. 3-11) using the whipping or U motion with the tip and welding rod at opposite sides of the U legs as the weld progresses.

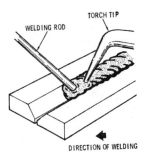

Forehand welding method. In this method, the welding rod moves ahead of the torch tip.

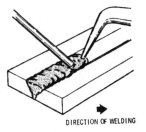

Backhand welding method. In this method, the torch moves ahead of the welding rod.

Fig. 3-10. Forehand and backhand welding techniques.

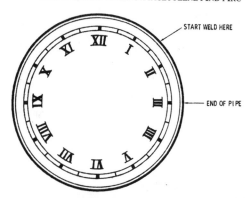

Fig. 3-11. Starting a weld at the 1:30 position.

Making a Position Weld

A position weld is one that is made on a pipe that is stationary, not rolled during the welding process. A position weld is the most difficult weld to make, yet job conditions often require that this type of weld be made. If the technique for making position welds is learned in the beginning, the welder can tackle any job with confidence. In making position welds, the welder must continually change torch angle and welding rod angle as the weld progresses around the pipe. The correct angles of the rod and torch in the various positions are shown in Fig. 3-12.

The pipe will be held stationary in a horizontal position to make the practice weld. Tack weld a short piece, 18 or 20 in. long, of $1/2"$ or $5/8"$ steel rod to one end of the pipe; then place the other end of the rod in a vise (Fig. 3-13).

When making this weld start at the bottom (A in Fig. 3-12) and weld to the top (I) on one side. Then go back to the bottom and weld to the top on the other side. Heat the welding rod about 8" above one end and make a 90° bend. Play the flame on both pieces of pipe at A (Fig. 3-12) until the metal melts and starts to run down into the gap. Insert the short end of the welding rod at a 45° angle into the melted pool and build up the pool to the top of the metal. The pool will begin to run ahead of the flame, and the metal should be white hot and melting as the pool meets it. Make this practice weld in one pass, moving the welding rod and the torch tip in U-shape motions with the end of the rod and the tip at opposite ends of the U. This is a one-pass weld with the gap filled to slightly above the top as the weld progresses. The tip and rod angles will

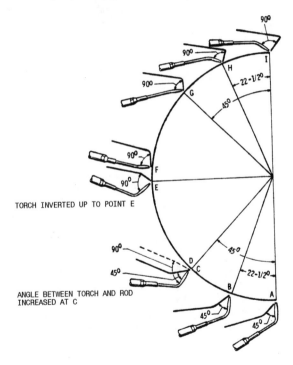

TORCH INVERTED UP TO POINT E

ANGLE BETWEEN TORCH AND ROD
INCREASED AT C

Fig. 3-12. Holding the torch and rod correctly for a position weld.

ROD TACKED TO PIPE

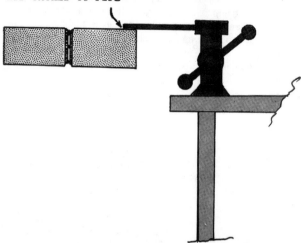

Fig. 3-13. Pipe supported in a vise, ready for position weld.

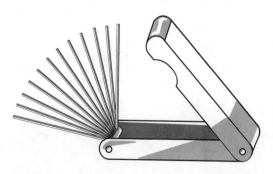

Courtesy National Cylinder Gas Div. of Chomtron Corp.

Fig. 3-14. A welding tip cleaner.

change as shown in Fig. 3-12 as the different positions from A to I on each side are reached. As you learn to make position welds, you will 'get the feel' for welding. After a practice position weld has been made, the weld can be cut out, and the pipe can be used again for practice.

TABLE 3-1 shows the correct tip sizes, oxygen, and acetylene pressures for various metal thicknesses.

The orifices in the cutting head should be cleaned often, using a tip cleaner, Fig. 3-14.

Table 3-1. Oxyacetylene Welding Tip Data

Tip Size	Drill Size	Oxygen Pressure psig		Acetylene Pressure psig		Acetylene Consumption CFH*		Metal Thickness
		Min.	Max.	Min.	Max.	Min.	Max.	
000	75	1/4	2	1/2	2	1/2	3	up to 1/32"
00	70	1	2	1	2	1	4	1/64"—3/64"
0	65	1	3	1	3	2	6	1/32"—8/64"
1	60	1	4	1	4	4	8	3/64"—3/32"
2	56	2	5	2	5	7	13	1/16"—1/8"
3	53	3	7	3	7	8	36	1/8"—3/16"
4	49	4	10	4	10	10	41	3/16"—1/4"
5	43	5	12	5	15	15	59	1/4"—1/2"
6	36	6	14	6	15	55	127	1/2"—3/4"
7	30	7	16	7	15	78	152	3/4"—1 1/4"
8	29	9	19	8	15	81	160	1 1/4"—2"
9	28	10	20	9	15	90	166	2"—2 1/2"
10	27	11	22	10	15	100	169	2 1/2"—3"
11	26	13	24	11	15	106	175	3"—3 1/2"
12	25	14	28	12	15	111	211	3 1/2"—4"

*Oxygen consumption is 1.1 times the acetylene under neutral flame conditions. Gas consumption data is merely for rough estimating purposes. It will vary greatly with the material being welded and the particular skill of the operator. Pressures are approximate for hose length up to 25 ft. Increase for longer hose lengths about 1 psig per 25 feet. *Courtesy Victor Equipment Co.*

SAFETY PRECAUTIONS

Never allow oxygen under pressure to come in contact with oil, grease, dirt, or any type of organic material that may burn easily in high oxygen concentrations.

Never use an acetylene torch to blow off clothing.

Never use oxygen as a substitute for compressed air for testing purposes.

Never use acetylene at pressures in excess of 15 psig. (lbs. sq. in. gauge)

Always wear goggles of the correct shade when welding.

Always wear gloves when welding.

When welding or cutting on jacketed or hollow parts, make certain that the parts are vented.

In many areas, safety rules require that another person work alongside a welder. The other person acts as an observer and prevents the starting of fires outside the welder's line of vision.

SHIELDED METAL ARC WELDING (SMAW)

(Also called "stick" welding)

This section is designed to help the experienced welder perform the most difficult pipe weld, the 6G weld. Circumferential arc welding of pipe in the 6G position, pipe stationary at a 45° angle, is the most difficult task a welder faces. The American Welding Society designates welding positions from 1G through 6G. Most welding standards agree that a welder who can successfully complete a pipe test in the 6G position is qualified to weld pipe in all positions.

Learning basic welding techniques requires "hands on" training. Most welders learn to weld in apprenticeship programs, welding schools, or trade schools. Learning to make welds in all positions requires practicing correct procedures and techniques. The information presented here is designed to help the reader advance from lower qualification ratings to a 6G qualification.

Making a Position Weld

A 6G position weld must be made to join two pieces of 6" schedule 80 black steel pipe. The joint is in a pipe axis at a 45° angle as shown in Fig. 3-15. The ends to be joined are beveled at a 37° angle with a 3/32" root opening. The wall thickness of the pipe, .432", will require seven passes. The root pass will be made with 3/32" diameter E6010 electrodes. The hot, or second, pass is followed by two fill passes, and the weld is finished by three cover passes, all deposited with 1/8" diameter E7018 electrodes.

In preparation for the weld, the pipe must be cleaned by using a wire brush to remove scale, rust, or oil to a distance 1" back from the weld. The pipe can then be aligned in an angle or channel iron, as mentioned earlier, and tacked, leaving a 3/32" gap for the root pass.

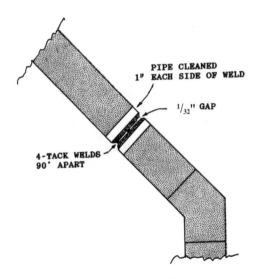

Fig. 3-15. Pipe in position for a 6G weld.

Setting the Welding Current

The weld will be made with the electrode positive and with the welding machine set to DC (direct current) at an acceptable range. Voltage and amperage will vary from one machine to another, therefore the welder should make a vertical test weld on a piece of scrap to determine the correct amperage setting.

Making the Tack Welds

The first step in joining is making the tack welds. Tack welds are part of the first pass. The tack welds should be $3/4$" in length and should be spaced 90° apart around the perimeter of the pipe. Fig. 3-15 shows the pipe in position for the weld. To assure a good tie-in to the root pass, the edges of tacks should be ground to a feather edge.

The Root Pass

The bead sequence for welding pipe in the 6G position divides the pipe circumference into quadrants and requires the welder to start each quarter circle weld at the farthest location from where the previous bead ended. This procedure ensures even heating and minimum distortion. To ensure good penetration of the root pass, the bead will be run uphill in all quadrants. Starting at A, the root pass of the first quadrant A to B is made. To avoid distortion of the pipe, the root pass in the second quadrant, D to C, is made next, followed by the third quadrant, A to D, and the fourth, B to C. Each quarter circle weld should start at the segment opposite to where the preceding weld segment ended, as seen in Fig. 3-16.

The arc should be struck about 1" in front of the point where the weld is to start. A long arc should be held until the electrode is dragged to the weld start point. At the starting point the arc should be shortened to the correct length, $1/16$ to $1/8$ in. This method allows the welder to find the starting point and allows time to establish proper shielding-gas flow and arc characteristics. When welding the root pass, the current should be adjusted to maintain a keyhole (molten metal puddle) about $1 1/2$ times the electrode diameter.

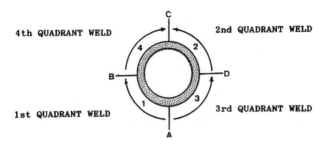

Fig. 3-16. Pipe divided into four quadrants.

If this keyhole is maintained, a good inside bead is formed with complete penetration. As the electrode moves forward in the joint, molten metal fills in behind the keyhole and forms the weld. To maintain the correct keyhole size, the welder adjusts arc current and weld travel speed. If the current is set too low, the keyhole becomes too large, causing internal undercut or melt through. The welder should watch the molten puddle and the ridge of solidifying metal behind the arc. The ridge should form about 3/8 of an inch behind the electrode. If weld travel is too fast, a thin high-crowned bead will be formed.

Stepping Technique

When an E6010 electrode is used for the root pass, the welder steps the electrode in the joint using wrist action to move the electrode forward about 1 1/2 times its diameter, then returns it to the puddle. This process is repeated along the joint. The change in electrode angle and technique that occurs as the weld moves forward and upward must be gradual. Penetration will be incomplete if the welding is done with the electrode (rod) at too low an angle or at excessive travel speed.

Penetration can be increased by increasing the rod angle or decreasing travel speed. To maintain uniform and proper root penetration, the electrode should be held at a 90° angle to the pipe circumference when starting the weld, Fig. 3-17. As the electrode is moved forward in making the root pass, the welder will change the angle of the electrode slightly, plus or minus 5°, to maintain the "puddle" in front of or behind the arc. The change in angle is called a "leading" angle or a "trailing angle."

When welding above the 4-o'clock and 8-o'clock positions, the electrode should not exceed a 5° trailing angle. Any time the root pass is stopped, the welder should push the electrode through the keyhole about 1/2 inch, extinguish the arc, and withdraw the electrode. This procedure helps maintain full root penetration at tie-ins. After welding each segment, a grinder should be used to feather the ends of the bead. When the final quadrant of the root pass is completed at the top of the pass, the weld must be cleaned. A chipping hammer and a wire brush should be used to remove the slag. Using a die grinder to form the root pass into a U shape will help insure that on subsequent passes the electrode will touch bottom all the way around the joint. After grinding, the sandy residue left by the grinder must be removed with a wire brush. If this residue is not removed, it will mix with the molten metal on the next pass and contaminate the weld.

The Hot Pass

The second or "hot pass" will require a current setting slightly higher than was used for the root pass. The hot pass must have enough current to burn out impurities in the root pass, while at the same time not so high as to cause the electrode coating to break down or burn through the root pass. A side-to-side weave motion not exceeding three times the welding rod diameter will provide full bead width.

At the start of the root pass, the electrode was held at a 90° angle to the weld; for the hot pass and all subsequent passes, the electrode should be angled slightly to allow for complete penetration into the previous pass and to the pipe wall. Slag must be removed from every pass, using a wire brush and chipping hammer. If there is a visible flaw, the flawed area should be ground out and the residue from the grinder removed.

The Fill Passes

When the two fill passes are made, the first fill pass should be made on the low side of the joint and the second pass on the high side. A slight oscillating or up-and-down motion of the electrode allows the molten weld metal to penetrate into both the previous pass and the side of the pipe wall. Each pass should be carefully checked for undercut and weld porosity.

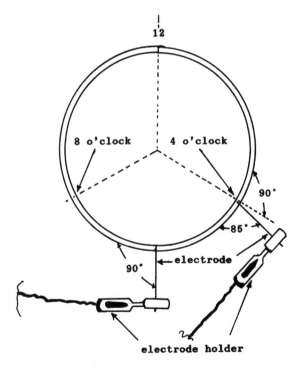

Fig. 3-17. Electrode angle changes as weld progresses.

The Cover Passes

Each cover pass begins on the low side of the joint using the oscillating technique as was used on the fill passes. Each quadrant is welded until the weld surrounds the pipe. After the first pass is cleaned, the second pass is made, penetrating equally into the first cover pass and into the fill pass below. After the second cover pass is cleaned, the third cover pass is made. This pass should fill the remaining gap. If the entire gap cannot be filled with the third pass another pass will be needed.

When grinding to remove high spots on the finished weld, the abrasive should be run *across* the weld, not along the weld. In an X-ray, parallel grinding marks could be mistaken for flaws in the weld.

SAFETY PRECAUTIONS

A welder and anyone watching a welding procedure must use a welding helmet equipped with the proper shaded lens for adequate eye protection. Shaded lenses in the range of 12 to 14 are suggested.

Before operating a welding machine, check to be certain that the machine is properly grounded.

The area where welding is being done should be so enclosed that the arc is not visible outside this area.

Heavy leather gauntlet type gloves should be worn while welding. Never weld while wearing wet gloves or wet shoes.

Keep the working area clean; pick up electrode stubs, scrap metal, etc.

Never weld on closed containers or on containers that have contained combustible materials.

Welding inside tanks, boilers, or other confined spaces requires special procedures such as a hose mask or air-supplied hood. Oxygen depletion when working in a confined space or vessel can have fatal results.

Welding or cutting galvanized materials is very dangerous because of the fumes released by burning zinc. High velocity fans should be so placed that the fumes are blown away from the welder and the area is well ventilated to protect other personnel in the room, building, or vicinity.

More complete information on health protection and ventilation can be found in the American Standard Z. 49.1 "Safety in Welding and Cutting." This document is available from:

The American Welding Society, P.O. Box 351040, Miami, FL 33135.

CHAPTER 4

AUTOMATIC FIRE PROTECTION SYSTEMS

Sprinkler fitting is a very specialized branch of the piping trade. Sprinkler fitters serve an apprenticeship during which they learn the various codes and standards that apply to fire protection installations. The tools and machinery used by sprinkler fitters are common to other branches of the piping trades. The scope of work involved includes installing water mains to serve sprinkler systems; installation of piping from mains to and including individual piping systems; installation and servicing of valves, air compressors, and foam generators; and regular maintenance, testing, and servicing of various types of automatic fire protection equipment.

A sprinkler system designed for fire protection is an integrated system of underground and overhead piping which includes one or more automatically controlled water supplies. The portion of the sprinkler system located above ground is a network of specially sized or hydraulically designed piping installed in a building, structure or area, generally overhead, to which sprinkler heads are attached in a systematic pattern. The valve controlling each system riser is located in the riser or its supply piping. Each sprinkler riser includes a device for actuating an alarm if the system is activated.

Normally activated by heat from a fire, the system discharges water over the fire area through sprinkler heads. Three typical sprinkler heads are shown in Fig. 4-1.

Sprinkler systems can be classified into two main types:

1. Wet-pipe systems.
2. Dry-pipe systems.

Courtesy The Viking Co.

Fig. 4-1. Three types of sprinkler heads.

Further variations of these systems are:

1. Pre-action systems.
2. Deluge systems.
3. Combined dry-pipe and pre-action systems.

WET-PIPE SYSTEM

A wet-pipe sprinkler system is fixed fire protection using piping filled with pressurized water and activated by fusible sprinklers for the control of fire.

A wet-pipe sprinkler system may be installed in any structure not subject to freezing temperatures. The system will automatically protect the structure, contents, and/or personnel from loss or harm due to fire. However, the structure must be substantial enough to support the piping system filled with water.

Operation of a Wet-Pipe System

When a fire occurs, the heat produced will fuse a sprinkler and cause water to flow. The alarm valve clapper is opened by the flow and allows pressurized water to fill the retarding chamber. The flow overcomes the retarding chamber's small capacity drain and fills the alarm line. This in turn closes the pressure switch, sounds an electric alarm, and activates the mechanical water motor alarm. If a water-flow indicator is used in the system piping, it also is activated by the water flow. The paddle, which normally lies motionless inside the pipe, is forced up, activating the pneumatic time-delay mechanism. This in turn, energizes a micro switch, causing an alarm to sound as long as water is flowing through the system. The water will flow until it is shut off manually. Components of a wet-pipe system are shown in Fig. 4-2.

DRY-PIPE SYSTEM

A typical application of a dry-pipe system is in a structure that is not heated and is subject to below-freezing temperatures. A dry-pipe sprinkler system is a fire protection system that uses water as an extinguishing agent, but differs from a wet-pipe system in that the piping from the dry-pipe valve to the fusible sprinkler heads is filled with pressurized air or nitrogen instead of water.

An air-check system is a small dry system which is directly connected to a wet-pipe system. The air check system uses a dry valve and an air supply but does not have a separate alarm. The alarm is provided by the main alarm valve. Although the system is of less weight than the wet-pipe system, the structure housing a dry-pipe system must, nevertheless, be substantial enough to support the system piping when the system is filled with water.

Operation of a Dry-Pipe System

The components of a dry-pipe system are shown in Fig. 4-3. These components may vary somewhat due to the application of different sets of standards. The system shown in Fig. 4-3 is only one possible arrangement of a dry-pipe system. Additional

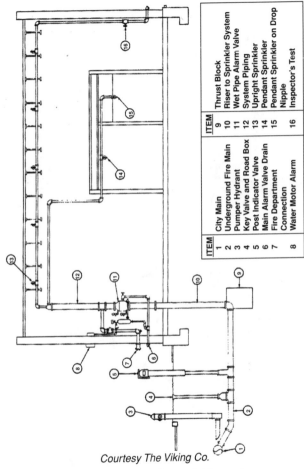

ITEM	
1	City Main
2	Underground Fire Main
3	Pumper Hydrant
4	Key Valve and Road Box
5	Post Indicator Valve
6	Main Alarm Valve Drain
7	Fire Department Connection
8	Water Motor Alarm
9	Thrust Block
10	Riser to Sprinkler System
11	Wet Pipe Alarm Valve
12	System Piping
13	Upright Sprinkler
14	Pendant Sprinkler
15	Pendant Sprinkler on Drop Nipple
16	Inspector's Test

Courtesy The Viking Co.

Fig. 4-2. Components of a wet-pipe sprinkler system.

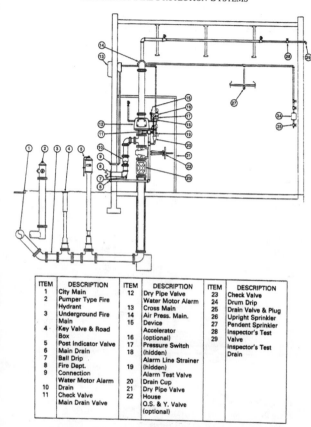

ITEM	DESCRIPTION	ITEM	DESCRIPTION	ITEM	DESCRIPTION
1	City Main	12	Dry Pipe Valve	23	Check Valve
2	Pumper Type Fire		Water Motor Alarm	24	Drum Drip
	Hydrant	13	Cross Main	25	Drain Valve & Plug
3	Underground Fire	14	Air Press. Main.	26	Upright Sprinkler
	Main	15	Device	27	Pendent Sprinkler
4	Key Valve & Road		Accelerator	28	Inspector's Test
	Box		(optional)		Valve
5	Post Indicator Valve	16	Pressure Switch	29	Inspector's Test
6	Main Drain	17	(hidden)		Drain
7	Ball Drip	18	Alarm Line Strainer		
8	Fire Dept.		(hidden)		
9	Connection	19	Alarm Test Valve		
	Water Motor Alarm	20	Drain Cup		
10	Drain	21	Dry Pipe Valve		
11	Check Valve	22	House		
	Main Drain Valve		O.S. & Y. Valve		
			(optional)		

Courtesy The Viking Co.

Fig. 4-3. Components of a dry-pipe sprinkler system.

components of a dry-pipe system are an adequate supply of water taken from a city main, elevated storage tank, or ground storage reservoir with an adequate pumping system.

Underground System

1. Piping. Cast iron, ductile iron, cement asbestos, and (where permitted) PVC Schedule 40.
2. Control valves, Post Indicator Valves. (PIV).
3. Valve Pit. Usually required when multiple sprinkler systems are serviced from a common underground system taking water supply from a city main. Pit will contain: O.S. & Y (open stem and yoke) valves; check valves or detector check, fire department connection; $2^1/2" \times 2^1/2" \times 4"$ hose connection; 4" check valves with ball drip (backflow preventer).
4. Auxiliary Equipment. Fire hydrants with two $2^1/2" \times 2^1/2"$ outlets for hose line use and 4" outlet for fire pumper connection. (Outlets needed may vary depending on local fire department.) Structures for housing fire hose and equipment.

DELUGE SYSTEMS

A deluge system is an empty pipe system that is used in high hazard areas or in areas where fire may spread quickly. It can also be used to cool surfaces such as tanks, process piping lines, or transformers. In this type of application, open sprinklers or spray nozzles are employed for water distribution. The deluge valve is activated by a release system employing one or more of the following methods: manual, fixed temperature, rate of temperature rise, radiation, smoke or combustion gases, hazardous vapors, or temperature increase. Once a deluge system is activated, water or other extinguishing agents flow through all spray nozzles and/or sprinkler heads simultaneously. Two typical applications of deluge systems are flammable liquid loading facilities and aircraft hangars.

Flammable Liquid Loading Facilities

Truck and rail loading facilities for flammable or combustible liquids routinely handle large volumes of potentially dangerous materials in complete safety. Vehicles may, however, collide or hit the loading structure, rupturing tanks or pipes. Hoses, nozzles, or valves that are in constant use may break or malfunction, and operator error is always a hazard. Any of these failures could result in a spill which, if ignited, could cause a disastrous fire. Since large volumes of fuel are present, it is likely that a fire would develop rapidly and endanger personnel, vehicles, cargoes and the unprotected loading structure. The loading site is almost always curbed or otherwise contained since it is not usually permissible to wash the fire to another area. For that reason the fire must be quickly extinguished and not allowed to re-flash.

Control Strategy

A widely used and economical way to handle the hazard is a foam-water deluge sprinkler system. The structure and the upper portion of the vehicles are kept cool by a water spray, but because water alone is not effective in the extinguishment of bouyant burning liquids, a foaming agent must be introduced into the water.

Because a fire may be expected to develop very quickly, combination rate-of-rise and fixed temperature detectors are normally employed. Back-up manual activation should also be provided.

Pipe lines carrying flammable liquid or fuel to the loading area may be equipped with shut-off valves which are automatically closed when the fire detectors operate. Pumps associated with this system may also be interlocked to shut off.

When the detection system operates, the deluge valve opens and supplies water to the sprinkler piping. Aqueous film-forming foam (AFFF) concentrate is introduced into the water by the proportioner and is mixed in the sprinkler piping. Unsealed upright sprinklers provide for the structure, the top of the vehicles, and the surrounding area. Unsealed sprinklers located on the lower structure deliver foam to the underside of the vehicle and over the ground below. Hand-held hose lines controlled by manual valves are also fed from the sprinkler piping. Water flow alarms sound when the sprinkler piping is flooded. Flammable liquid supply may be stopped by pump shutdown or by an automatic shutoff valve actuated by a pressure switch in the sprinkler piping. The components of a fire protection system for liquid loading facilities is shown in Fig. 4-4.

Protein foam may be used in place of AFFF. If protein foam is used, special foam-water sprinklers must replace standard open sprinklers. If large quantities of water and foam concentrate are required, pumps which are started by a signal from the detection system are usually employed for both liquids. Flow-control valves with downstream pressure regulators may be used in place of deluge valves to conserve water and foam concentrate.

In the event that release is in a freezing area, a pneumatic release system must be used and a suitable air supply provided. Electrical fixed-temperature and rate-of-temperature rise detection may be used in place of mechanical detection, but such detectors may be required to meet hazardous area standards. Infrared or ultraviolet detection may also be used, but consideration must be given to the increased potential for false trips.

Aircraft Hangar Fire Protection

Today's huge aircraft hangar represents a large and concentrated loss potential. The primary hazard is the ignition of a fuel spill involving an aircraft located in the hangar. An aircraft is extremely vulnerable to heat sources producing skin temperatures above 400°F (250°C). The hangar itself is susceptible to fire damage, since usual construction includes an unprotected steel-supported roof which will buckle and possibly collapse when exposed to high temperatures.

Control Strategy

Rapid control and extinguishment of fire is vital if loss is to be minimized. Even when it is not possible to save an aircraft, it is necessary to control and extinguish the fire to protect the hangar. Since buoyant flammable liquid is the principal fire hazard, extinguishing systems employing a foam-water agent are the preferred means of protection. The aircraft and the building structure are cooled by the water while the burning fuel is smothered by the foam. Water alone generally means higher density requirements, resulting in higher costs and lower efficiency.

It is not practical to mount fixed fire protection directly under the aircraft where fire is most likely to occur. Therefore, three types of extinguishing systems are normally provided for complete coverage, all employing foam. First, a monitor nozzle system, automatically activated by high-sensitivity optical detectors, is used to protect the underside of the aircraft which is shielded from standard foam-water sprinkler protection installed at the hangar roof. Second, the roof sprinkler system, activated by rate-of-rise detectors, provides general area protection, usually deluge. This foam-water system provides building protection in the event that the monitor nozzle system fails to control a fire. Third, hand-held hose lines with foam nozzles are provided for manual fire fighting operations. Because building protection is provided by a deluge system, large amounts of water and foam concentrate are required. These must be properly managed to avoid exhausting the supply. Pumping facilities are nearly always required to meet high volume demands.

OPERATION OF A FOAM-WATER DELUGE SYSTEM

Initial detection is normally accomplished by the monitor detection system employing either infrared or ultraviolet detectors. These systems are capable of detecting a fire in its early stages. However, false trips with such detection systems are a possibility and must be seriously considered. Manual activation of the monitor system is also provided.

In a typical system actuation, a signal from the monitor detector is received by the release control panel (which may be interlocked to the foam and water concentrate supply pump starters). This signal opens both flow control valves supplying water and foam to the monitor nozzles and sounds an alarm. Foam concentrate is proportioned into the water and mixed in the supply piping. The flow control valve operates both as a deluge valve and as a downstream pressure regulating valve to insure the correct proportion of water and foam concentrate regardless of varying supply pressure. The monitor nozzles move in a programmed path to cover the aircraft, the under-aircraft areas, and areas that might contain a fuel spill. Hose lines supplied by individual manual valves are located at convenient places for manual fire fighting.

In the event the monitor nozzles do not control the fire, the temperature will increase at the roof. This condition will trigger a release which is activated by either fixed-temperature or rate-of-temperature rise conditions causing the flow control valve to open. Unsealed, upright sprinklers then distribute the foam over the general area. Operation of the roof system will trip the monitor system in the event that it has not already functioned. That operation may also be designed to actuate systems in adjacent areas. A schematic drawing of an aircraft hangar fire protection system is shown in Fig. 4-5.

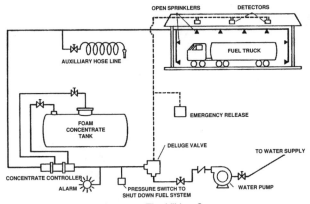

Courtesy The Viking Co.

Fig. 4-4. Schematic drawing of a foam deluge system used to protect liquid loading facilities.

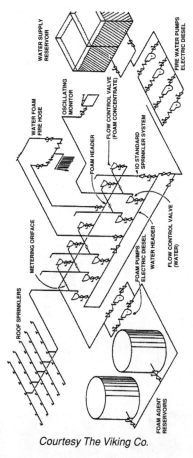

Courtesy The Viking Co.

Fig. 4-5. Schematic drawing of an aircraft hangar fire protection system.

Courtesy The Viking Co.

Fig. 4-6. A double-check backflow preventer.

Protein or fluoroprotein foam may be used in place of AFFF. If such foam is used, special foam-water sprinklers must also be used in place of standard open sprinklers. In the event that the release is in a freezing area, pneumatic operation must be used and a suitable air supply be provided.

Electrical fixed-temperature and rate-of-temperature-rise detection utilizing a release control panel may be used in place of mechanical detection at the hangar roof.

Standard practice in most areas is to use Schedule 40 black steel pipe and threaded black cast iron fittings for mains and branch piping. 2 1/2" I.D. (inside diameter) and larger valves and fittings are usually flanged and joined to threaded pipe by companion flanges.

Backflow Preventers

When fire protection systems are connected to potable water supplies, check valves are required in the system piping to prevent backflow of potentially polluted water into the potable water system. When fire department pumper trucks arrive at a fire, they immediately connect a hose to the nearest fire hydrant. A hose may, if needed, be connected from the truck's pump to the fire department Siamese connection (Figs 4-2 and 4-3). Water is drawn from the hydrant, and if higher pressure is needed, the pumper truck can pump water into the fire protection system piping. The pumper truck is capable of building pressure in excess of water main pressure and, if the backflow preventer were not installed, could force potentially polluted water into the potable water main.

A double-check backflow preventer is shown in Fig. 4-6.

Black steel pipe, grooved or threaded with grooved or threaded fittings, is usually used for sprinkler piping. More information on grooved piping will be found in Chapter 9, Grooved Piping Systems.

CHAPTER 5

STEAM HEATING SYSTEMS

A typical steam boiler with all essential controls and safety devices is shown in Fig. 5-1. To be efficient in producing steam, a boiler must be clean; oils, greases, pipe dopes or other contaminating materials present in the boiler and piping must be purged. To accomplish this, the boiler must be filled to normal water level to which caustic soda or other suitable cleaning agent has been added. The boiler must then be brought up to steaming temperature, the drain connection opened, and the dirty water piped to a suitable drain. Water must be added during this process to replace the water drained out. The time needed to purge the boiler and piping depends on the boiler size.

All steam heating systems are one of two basic types: one-pipe or two-pipe systems. A one-pipe system uses one pipe to deliver steam to a unit and to return the condensate from the unit. There are variations of these two systems. In a parallel-flow system, both steam and condensate flow in the same direction (Fig. 5-2). When the steam flows in one direction and the condensate returns against the steam flow, the system is then called a counter-flow system (Fig. 5-3.) The heating unit has only one connection, and steam enters the unit while condensate is returning through the same connection. Correct pitch or slope of piping for proper operation of a gravity flow system is essential to ensure the flow of steam, air, and condensate. The correct pitch or slope for a gravity flow system is not less than one (1) inch in 20 feet. A one-pipe system is very dependable and initial installation cost is low. Air vents are used on each individual heating unit and at the ends of steam mains.

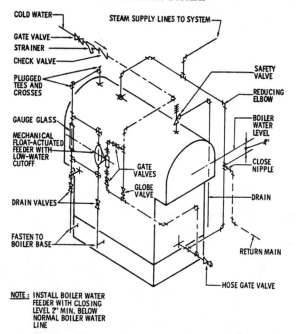

Fig. 5-1. A steam boiler with all essential controls.

Steam must enter the heating unit at the bottom inlet tapping of the unit, and because steam must enter and condensate return through the valve at the same time, an angle pattern valve should be used at this point. A straightway horizontal valve would offer an obstruction to the free flow of steam and condensate. A radiator supply valve on a one-pipe system can not be used as a throttling valve; the valve must be either full ON or full OFF. If valves on this type system were used as throttling valves, condensate could not return to the boiler and noise would be created. The correct valve for use in a one-pipe steam system is shown in Fig. 5-4. Air present in the piping of a one-pipe system must be eliminated in order for steam to flow. An end-of-main vent should be installed

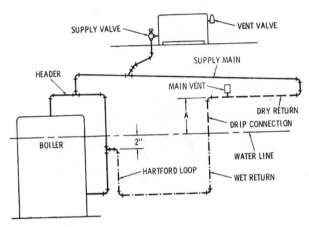

Fig. 5-2. A one-pipe gravity parallel-flow steam system.

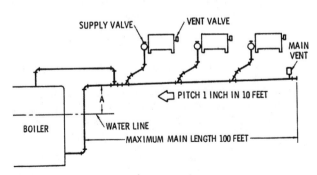

Fig. 5-3. A one-pipe gravity counter-flow steam system.

Fig. 5-4. The correct radiator valve for a one-pipe steam system.

near, but not at, the end of the steam main. The right location for this vent is shown in Fig. 5-5(A). If this valve is not installed correctly, the float can be damaged by water surge. A vacuum type air vent prevents air from entering the heating unit after venting has occurred.

Air must be vented from radiators to allow steam to enter. As a result, an air vent must be installed in the vent tapping on the side opposite the supply valve. There are two types of air vents: the open (non-vacuum) type, which has a single non-adjustable port, and the adjustable-port vacuum vent used for proportional venting. Proportional venting is best; it allows steam to enter all radiators at the same time if vents are correctly adjusted. Air vents function

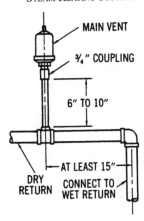

Fig. 5-5(A). The correct location for an end-of-main vent.

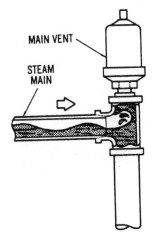

Fig. 5-5(B). The wrong location for an end-of-main vent.

by opening to allow steam to enter and by closing when steam or water contacts the vent. An adjustable-port air vent is shown in Fig. 5-6.

Wet Returns—Dry Returns

At the end of the steam main, steam returns to water and becomes condensate. This condensate is joined by the condensate formed in various units of radiation and returns to the boiler through the return piping. The section of piping between the end of the supply main and the end-of-main vent is called the *dry return*. The dry return also carries steam and air. It is that portion of the return main which is located above the boiler water line.

That portion of the return main which is below the boiler water line is called a *wet return*, shown in Fig. 5-17. A wet return is also shown in Fig. 5-7; a *dry return* is not shown here because the dimensions shown are used only in designing a steam heating system. In both the wet and dry return systems, dimension "A" in Fig. 5-7 must be maintained.

Hartford Loop

A Hartford loop (Fig. 5-2) is a pressure balancing loop which introduces full boiler pressure on the *return* side of the boiler. Without this loop, *reversed circulation* could occur, allowing water to leave the boiler via the return piping. It is the only safe method of preventing reversed circulation. The Hartford loop must be the full size of the return main, and the horizontal nipple at the return connection point should be 2 inches below the boiler water line. A close nipple should be used to construct the Hartford loop. If a longer nipple is used, water hammer noise will be created. Water hammer is a wave transmitted through a pipe filled or partially filled with water. The rapid passage of steam over the surface of the water causes a wave to form. When this wave slams against another pocket of water or the float of a vent valve, the hammering noise is created.

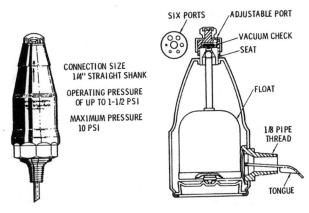

CONNECTION SIZE
1/4" STRAIGHT SHANK

OPERATING PRESSURE
OF UP TO 1-1/2 PSI

MAXIMUM PRESSURE
10 PSI

SIX PORTS | ADJUSTABLE PORT
VACUUM CHECK
SEAT
FLOAT
1/8 PIPE
THREAD
TONGUE

Fig. 5-6. An adjustable-port air vent.

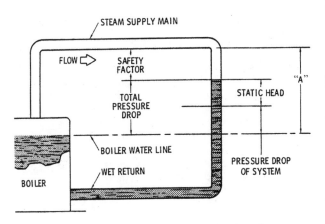

STEAM SUPPLY MAIN
FLOW ⇨ SAFETY FACTOR
"A"
TOTAL PRESSURE DROP
STATIC HEAD
BOILER WATER LINE
WET RETURN
BOILER
PRESSURE DROP OF SYSTEM

Fig. 5-7. Measuring pressure drop, static head, and safety factors.

System Pressures—Designing a Steam Heating System

A small system having a total heat loss of *not more than 100,000 Btu/hr* is sized on the basis of $1/8$ psig (lbs. sq. in. gauge). The three distances for the pressure drop, the static head, and the safety factor in Fig. 5-7 will be:

> Pressure drop of system ($1/8$ psig) = $3^1/2$ inches of water
>
> Static head (friction of wet return) = $3^1/2$ inches of water
>
> Safety factor (twice the static head) = 7 inches of water
>
> Total distance = 14 inches of water

It is standard practice to make distance "A" in small systems not less than 18 inches.

For a larger system, assume that the piping is sized for a pressure drop of $1/2$ psi. The three distances shown in Fig. 5-7 would then be:

> Pressure drop of system ($1/2$ psig) = 14 inches of water
>
> Static head (friction of wet return) = 4 inches of water
>
> Safety factor (twice the static head) = 8 inches of water
>
> Total distance = 26 inches of water

It is standard practice to make minimum dimension "A" for a system based on $1/2$ inch pressure drop not less than 28 inches.

To explain the above figures:

Fig. 5-7 shows that water in the wet return is really in an inverted siphon with the boiler steam pressure on top of the water at the boiler end and steam main pressure on top of the water at the other end. The difference between these two pressures is the *pressure drop* in the system. The pressure drop is the loss due to friction of the steam passing from the boiler to the far end of the steam main. The water at the far end will rise sufficiently to overcome this difference in order to balance the pressures, and it will rise enough, usually about three inches, to produce a flow through the return into the boiler.

If a one-pipe system is designed for a total pressure drop of $1/2$ psi and a Hartford loop is used on the return, the rise in the water level at the far end of the return, due to the difference in steam pressure, will be $1/2$ of 28 inches or 14 inches. Adding three inches to this for the flow through the return main and 6 inches for a safety factor gives 23 inches as the distance by which the *bottom of the lowest part of the steam main and all heating units* must be above the boiler water level. Higher pressure drops would increase the distance accordingly.

More on Parallel Flow Systems

A one-pipe parallel flow up-feed system is designed for buildings having more than one floor. With this system, steam is distributed to various units of radiation through a basement main from which up-feed risers feed the second and third floors. The up-feed risers are dripped back to the return main. The risers should be connected as shown in Fig. 5-8 to keep the horizontal main free of condensate accumulation and to assure the unobstructed flow of steam. A parallel flow down-feed system has the distribution main installed above the heating units (Fig. 5-9). With this type system, steam and condensate flow in the same direction in the down-feed risers. The main steam supply riser should be installed

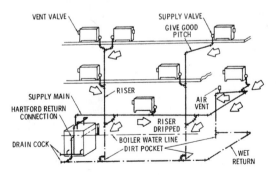

Fig. 5-8. A one-pipe parallel-flow up-feed system.

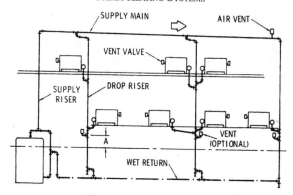

Fig. 5-9. A one-pipe parallel-flow down-feed system.

directly from the boiler to the overhead supply mains. The down-feed runout connections are taken from the bottom of the horizontal supply main to assure the least accumulation of condensate in the main.

Condensate Pumps

When there is insufficient height to maintain minimum dimension "A," Fig. 5-7, a condensate pump must be used to return condensate to the boiler. When a condensate pump is used, the boiler pressure, the end of the steam main pressure, and the boiler water line elevation have no bearing on the height of the end of the steam main *as long as it is above the maximum water level in the condensate pump receiver.* The return piping must be pitched to eliminate pockets which would trap air and prevent gravity flow of condensate to the receiver. The pump discharge is connected directly to the boiler return opening *without the use of a Hartford loop.* A Hartford loop connection can cause noise when used with a pumped discharge of condensate. Correct installation of an above-ground condensate receiver and pump is shown in Fig. 5-10. The correct way to install an underground receiver and pump is shown in Fig. 5-11.

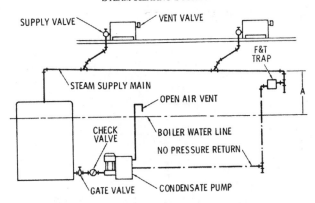

Fig. 5-10. A condensate receiver and pump installed above ground.

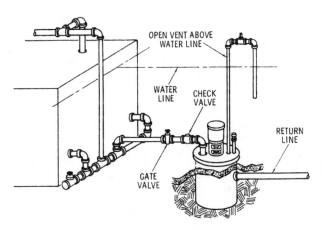

Fig. 5-11. A condensate receiver and pump installed underground.

Solving Problems on the Job

Fig. 5-12 shows the correct way to run steam and condensate piping around a beam.

Fig. 5-13 illustrates the correct way to run steam and condensate piping around a doorway.

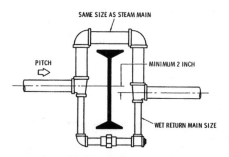

Fig. 5-12. Correct way to pipe steam and condensate around a beam.

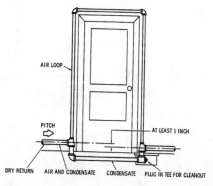

Fig. 5-13. Correct way to pipe steam and condensate around a doorway.

Fig. 5-14 demonstrates that a radiator must be level to prevent trapped condensate.

Fig. 5-15 exemplifies how the use of an eccentric reducer coupling will prevent trapping of condensate.

Fig. 5-16 explains how installing hangers to take sag out of a steam main, thus allowing condensate to flow, will stop noise.

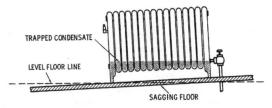

Fig. 5-14. A radiator must be level to prevent trapped condensate.

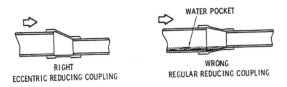

Fig. 5-15. An eccentric reducer prevents trapped condensate.

Fig. 5-16. Installing hangers to take sag out of piping will stop noise.

Two-Pipe Heating Systems

A two-pipe heating system is a system in which the heating units have two connections, one for the steam supply the other for condensate return. The return main begins at the discharge of a float and thermostatic trap as shown in Fig. 5-17. This trap is sized to handle the entire maximum condensate load of a single main or the connected load of individual mains. Some of the components used in one-pipe systems are also used in two-pipe systems. Two-pipe systems are designed to operate at pressures ranging from sub-atmospheric (vacuum) to high pressure. Condensate may be returned to the boiler by gravity flow or by the use of pumps or other mechanical return devices.

Thermostatic steam traps (Fig. 5-18) are the most commonly used traps in two-pipe steam heating systems. Thermostatic traps open in response to pressure as well as temperature in order to discharge condensate and air. When steam reaches the thermostatic element, the trap closes to prevent the discharge of steam into the return piping.

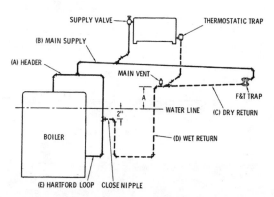

Fig. 5-17. The main components in a two-pipe gravity return system.

A float and thermostatic trap (Fig. 5-19) is opened by condensate collecting in the trap and raising the float. As the float is raised, it opens the discharge port allowing condensate to enter the return piping. When air or condensate is present in the trap at a temperature below its designed closing pressure, the thermostatic air by-pass remains open. The thermostatic air by-pass closes as steam enters the trap.

Several different types of traps are used in two-pipe steam heating systems. Some include float and thermostatic traps, float traps, inverted bucket traps, and open bucket traps.

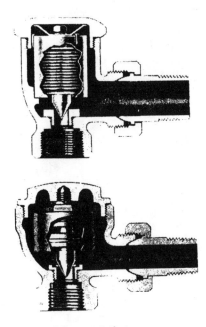

Fig. 5-18. Two types of thermostatic traps.

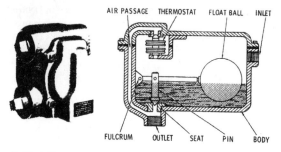

Fig. 5-19. A float and thermostatic trap.

Supply valves used with a two-pipe system are globe valves, either angle or straight type. Modulating (adjustable flow) or non-modulating type can be used. A packless type valve (Fig. 5-20) is best suited for use in two-pipe vacuum systems because the construction of the valve prevents air leakage into the system and steam leakage from the system.

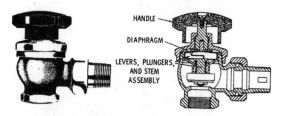

Fig. 5-20. A packless type valve.

Boiler Water Levels

It is very important that the water level in a steam boiler be maintained within correct limits at all times. If the water level is too high, inefficient operation results. If too low, the boiler may be permanently damaged or an explosion could occur. A boiler low-water cutoff or a combination boiler-water feeder and low-water cutoff (Fig. 5-21) is designed to automatically shut down power to a boiler in the event of a low water condition. Try cocks (Fig. 5-22) are installed on steam boilers at the safe high water level and at the safe low water level to provide for manual checking of the water level. The correct way to open a try cock is shown in Fig. 5-23.

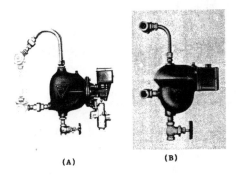

(A) (B)

(A) A combination boiler-water feeder and low-water cutoff.
(B) A boiler low-water cutoff.

Fig. 5-21. A low-water cutoff and a combination water feeder and low-water cutoff.

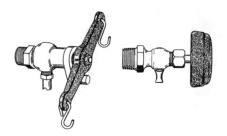

Fig. 5-22. Two types of boiler try cocks.

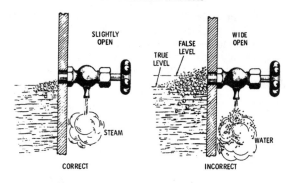

Fig. 5-23. Correct and incorrect way of testing the water level in a steam boiler.

INTERESTING FACTS ABOUT STEAM

A cubic inch of water evaporated under ordinary atmospheric pressure (14.7 psig) will be converted into approximately one cubic foot of steam. This cubic foot of steam exerts a mechanical force equal to lifting 1,955 lbs. one foot high.

The specific gravity of steam at atmospheric pressure is .462 that of air at 32°F and .0006 that of water at the same temperature. Therefore, 28.21 cubic feet of steam at atmospheric pressure weighs one pound, and 12.387 cubic feet of air weighs one pound.

Each nominal horsepower of boilers requires from four to six gallons of water per hour.

Good boilers will evaporate from 10 to 12 pounds of water per pound of coal.

One square foot of grate surface will consume from 10 to 12 pounds of hard coal or from 18 to 20 pounds of soft coal per hour with natural draft. With forced drafts these amounts can be doubled.

In calculating the horsepower of boilers, allow about 11½ square feet of heating surface per horsepower.

The standard unit of the horsepower of boilers (1915 Power Test Code, A.S.M.E.) is: One boiler horsepower is equivalent to the evaporation, from 212°F. feed water, of 34½ pounds of water into dry saturated steam at 212°.

Steam at a given temperature is said to be saturated when it is of maximum density for that temperature. Steam in contact with water is saturated steam.

Steam which has water in the form of small drops suspended in it is called "wet" or supersaturated steam. If wet steam is heated until all the water suspended in it is evaporated it is said to be "dry" steam.

If dry saturated steam is heated when not in contact with water, its temperature is raised and its density is diminished or its pressure is raised. The steam is then said to be "superheated."

LINEAR EXPANSION OF PIPING

Piping carrying steam or hot water will expand or lengthen in direct relation to the temperature of the steam or hot water. The formula for calculating the expansion distance is:

FORMULA: $E = constant \times (T - F)$
E = expansion in inches per hundred feet of pipe
F = starting temperature
T = final temperature

The constants per 100 ft. of pipe are:

Metal	Constant
Steel	.00804
Wrought Iron	.00816
Cast Iron	.00780
Copper-Brass	.01140

EXAMPLE: What is the expansion of 125 ft. of steel steam
             pipe at 10 psig pressure and a starting point of
             50 degrees?

$E = \text{constant} \times (T - F)$

Constant = 239.4 (see Appendix Table A-4 for boiling
point of water at 10 psig.)

$E = .00804 \times (239.4 - 50)$

$E = .00804 \times 189.4$

```
        .00804
      × 189.4
        003216
        007236
        006432
        00804
      1.522776
```

125 ft. of steel steam pipe will expand (lengthen) 1.522 inches
at 10 psig pressure with a starting point of 50° F.

CHAPTER 6

HOT-WATER HEATING SYSTEMS

A hot-water heating system consists of a coal, oil, or gas fired boiler, an expansion tank, boiler controls, piping, radiation, temperature controls, valves, air vents and circulating pumps if needed. Hot-water heating systems or combination heating and cooling systems are called "hydronic systems." Hot-water heating systems vary from simple one-pipe gravity systems to two-pipe reverse-return systems.

Boilers for hot-water heating systems are designed to operate at pressures not to exceed 30 psig. Two types of boilers are used: cast-iron sectional boilers and steel tubular boilers.

Cast iron boilers are made with front and back sections and one or more intermediate sections. The sections are assembled on the job and are joined together by tapered nipples at the top and bottom. When a sectional boiler is being assembled, extreme care must be taken to start the nipples straight. The sections are pulled together and held in place by steel rods inserted through an end section, then through intermediate sections, and through the other end section. Nuts and washers on each end of the rods must be tightened evenly as the sections are pulled together. The cold water inlet is at the bottom of the rear end section. The hot-water outlet is usually at the top of a center section. Tappings are provided in the sections for installation of controls. Modern boilers usually are equipped with a jacket installed after the boiler is installed. Older boilers may be insulated on the job and finished with a fabric coating.

Steel boilers are often delivered to the job site as a 'package' with an insulated jacket and many controls and accessories installed.

85

Large steel boilers are usually delivered as a bare unit, and all controls and accessories must be installed on the job. A refractory chamber must be built in the boiler for the firing method used. After all controls and accessories are installed, the boiler is insulated and a jacket is installed.

Expansion Tank

A compression tank, more often called an "expansion tank," (Fig. 6-1) must be installed in a hot-water heating system in order for the system to work properly. When water is heated, it expands, and in expanding, pressure is created. Water is virtually incompressible, so space must be provided to allow for expansion of the heated water. The expansion tank provides this space.

Although the expansion tank can be connected in several ways, the preferred method is to connect the tank to a tapping provided in an air separator. In normal operation, an expansion tank will be from $1/3$ to $1/2$ full of water. The air cushion in an expansion tank allows for the expansion and compression created when the water in the system is heated. If the tank is not airtight, air leaking from the tank will be displaced by water, leaving no space to allow for expansion-created pressure. An expansion tank filled with water is said to be "waterlogged." If an expansion tank becomes waterlogged without an air cushion, there is no room for expansion as the water in the system is heated. Pressure created by the

Fig. 6-1. A compression (expansion) tank used with hot-water heating systems.

expansion of the heated water will cause the relief valve to open and relieve the increased pressure. The expansion tank must be sized for the volume of water in the heating system. It is generally located in an area above the boiler.

A combination pressure, temperature, and altitude gauge must be installed on a hot-water boiler in order to check operating pressure and temperature.

Air Control

The air in an expansion tank should be the only air present in a hydronic system. But, water will absorb air, so a means must be provided to separate air present in the water and return it to the expansion tank. The air separator shown in Fig. 6-2. is made to be mounted on end- or side-outlet boilers. An in-line air separator is shown in Fig. 6-3. The flanged air separator shown in Fig. 6-4 is designed for large heating/cooling systems. The air separator shown in Fig. 6-5 is made for mounting on an expansion tank. It is designed to restrict the flow of water from the tank without restricting the flow of free air into the tank. The air separator (Fig. 6-6) is made for use in top-outlet boilers. Air is separated in the fitting and is directed to the expansion tank through the side outlet of the separator.

Fig. 6-2. Air separator for end- or side-outlet boilers.

Fig. 6-3. In-line air separator.

Fig. 6-4. An in-line air separator for large systems.

Fig. 6-5. A tank fitting that separates air from water.

Fig. 6-6. Air separator for top-outlet boilers.

(A) Manually operated air vent.

(B) Manual and/or automatic air vent.

(C) Automatic air vent.

Fig. 6-7. Three types of manual and automatic air vents.

Air in the piping system can cause noise and can also interfere with water circulation. When a new hot-water heating system is started up, or after a system has been drained, the water entering the system will force air to the high points. Air must be removed from high points in piping and from radiation to allow water to circulate through the system. This is accomplished by use of air vents installed on piping and radiation. Three types of air vents are shown in Fig. 6-7.

RELIEF VALVES

If an expansion tank becomes waterlogged or if a runaway firing condition should occur because of failure of controls, expansion and pressure could create a dangerous situation. An ASME/AGA temperature and pressure safety relief valve designed to open at 30 psi must be installed on every boiler. A typical safety relief valve for use on boilers is shown in Fig. 6-8.

System Operating Controls

Safety codes require that two aquastats be installed on every hot-water heating boiler. One aquastat serves as the operating control, controlling the boiler water temperature, the other as a

Fig. 6-8. An ASME boiler-water safety relief valve.

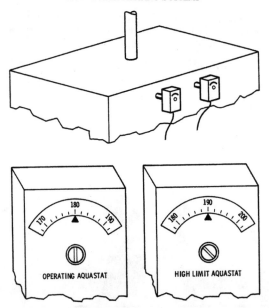

Fig. 6-9. Two aquastats must be used on hot-water boilers.

high limit control. The operating control (aquastat) is a normally open thermally-activated electrical switch that closes (makes contact) on temperature fall of the boiler water. When the switch closes, the gas or oil burner is energized, turning the burner on. When the boiler water reaches the set temperature, the aquastat switch opens, breaking electrical contact, and the burner is turned off.

The high limit control is a normally closed electrical switch that opens, breaking electrical contact if the boiler water temperature exceeds the set temperature of the high limit control. Normal practice is to set the high limit control 10 degrees above the operating control. Operating and high limit aquastats are shown in Fig. 6-9. Both the operating aquastat and the high limit aquastat are wired in series with the gas or oil burner controls. If either of the aquastats 'opens,' electrical contact with the gas or oil controls is broken and, providing the controls are wired correctly, the gas or oil burner will be shut off. The gas or oil controls can vary from simple solenoid valves to electronic and/or pneumatic controls which require technicians to maintain and service them.

Boiler operating temperatures can also be controlled by sensing devices installed outside a building. When this system is used, a rise in outside temperature will lower boiler water temperature; a drop in outside temperature will raise boiler water temperature.

Circulating Pumps

Circulating pumps are used to force circulation of water through the piping system. In one type of system, water circulates constantly, and temperatures of a floor, room, or area are controlled by electrically or pneumatically operated valves mounted on radiation and actuated by thermostats.

Temperatures in a room, floor, or area can also be controlled by a thermostat that electrically energizes a circulating pump, turning the pump on or off as temperatures rise or fall. This is called a 'zoned' system; each room, floor, or area is a zone with an individual run of piping with its own circulating pump.

An in-line circulating pump is shown in Fig. 6-10. This type pump is normally used on small heating systems. The pump shown in Fig. 6-11 is designed for larger systems. Many of these type pumps are high-velocity pumps that will force water through the system, pushing air ahead of the water, in some cases eliminating the need for air vents in the piping.

Boiler Water Feeders

Water is introduced into a heating system automatically by a valve of the type shown in Fig. 6-12. Given the fact that all water contains air, air in a heating system will rise to the highest point in the system. When a hot-water radiator does not get hot, it indicates that air is present in the radiator. When the air is bled off through the air vent, water enters the radiator replacing the discharged air. This causes a momentary drop in pressure, and water will enter the system through the automatic water feeder.

Hot-water heating systems are used for many purposes. Hot-water heat is used for radiant heat in floors and ceilings. Hot-water systems using anti-freeze solutions are used for melting snow and

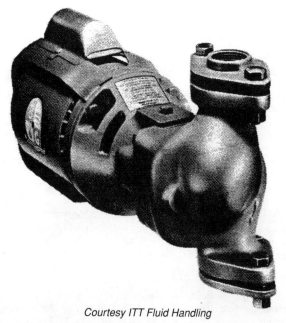

Courtesy ITT Fluid Handling

Fig. 6-10. An in-line circulating pump.

Courtesy ITT Fluid Handling

Fig. 6-11. A base-mounted circulating pump.

Fig. 6-12. An automatic boiler water feeder.

ice on sidewalks, parking garages, and other areas frequented by the public.

Whereas the water in a steam boiler must be free from grease or oil to allow the water to boil, oil or grease in a hot-water boiler does not affect its operating efficiency.

Hot-water heating systems where the water is heated by steam through a heat exchanger are often used in conjunction with a steam heating system. A word of warning here: The correct type of ASME/AGA rated relief valve must always be installed on this type of hot-water heating system.

Balancing a Hot-Water Heating System

A correctly sized hot-water heating system will heat every area in a building if the flow is evenly distributed to each piece of radiation. One problem often encountered is that the first three or four radiators on a piping run may heat while the last two or three remain cold. The first three or four radiators are taking all the heat from the supply, therefore the flow through these radiators must be restricted to allow hot water to reach the remaining radiators. Balancing cocks (Fig. 6-13) installed on the return side of each radiator and adjusted correctly will solve this problem.

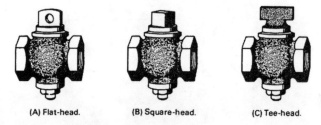

(A) Flat-head.   (B) Square-head.   (C) Tee-head.

Fig. 6-13. Three types of balancing cocks.

Thermostats

Although most thermostats are electric low voltage types, schools, public buildings, and many offices use pneumatic temperature control systems with pneumatically operated thermostats. It is apparent in every office building that the personal tastes of every worker vary as to the setting of thermostats. In public buildings it is common practice to install tamper proof covers over thermostats. A thermostat that is adjusted correctly will anticipate heating or cooling settings and regulate the temperatures accordingly.

THREE TYPES OF PIPING SYSTEMS FOR HOT-WATER HEAT

One-Pipe Forced-Circulation Systems

A typical one-pipe forced-circulation hot-water heating system is shown in Fig. 6-14. When this type system is used, there are several ways to force circulation through each piece of radiation. The two most common ways are to use reducing tees on the *inlet* and *outlet* connections to each radiator or to use *flow* fittings at the inlet connections.

Example Using Reducing Tees

Assuming the one-pipe system to be $1^1/4$" pipe, the *inlet* tee supplying the first radiator would be a $1^1/4$" × 1" × $^1/2$" tee. The pipe between the *inlet* tee and the *outlet* tee from the first radiator would be 1". The *outlet* tee would be a $1^1/4$" × 1" × $^1/2$" tee. The main would continue $1^1/4$" in size. The reduction in main size *between* the two tees forces circulation through the radiator. This system is shown in Fig. 6-14.

Example Using Monoflo Fittings

Again assuming the one-pipe system to be $1^1/4$" pipe, Monoflo fittings are made to be used as either *up-flow* or *down-flow* fittings. The built-in diverters in the Monoflo fittings serve as restrictors to

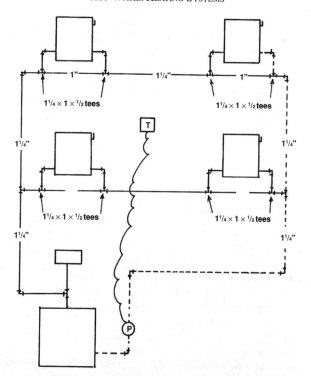

Fig. 6-14. A one-pipe system using reducing tees.

force flow through the radiators. As shown in Fig. 6-15, the Monoflo fittings should be installed with the "Ring" between the risers. On a one-fitting installation, the fitting would be installed on the return connection to the main with the arrow pointing in the direction of flow. When two Monoflow fittings are installed, the "ring" end should be between the risers.

Two-Pipe Direct-Return System

In a two-pipe direct-return system (Fig. 6-16) the supply main begins at the first radiator, the one closest to the boiler, and ends at the last radiator. The return main begins at the return side of the first radiator and connects to the return side of each radiator as it continues back to the return connection on the boiler. Each radiator in this type system has a different length supply and return, creating balancing problems. Balancing cocks (Fig. 6-13) will be needed on the return from each radiator to ensure even flow to each radiator in the system. Two aquastats (high and low limit controls) maintain boiler water temperature. A thermostat will operate the circulating pump to maintain set thermostat temperature.

Two-Pipe Reverse-Return Systems

Figure 6-17 shows a reverse-return system which has three zones. Each floor or area served by a supply-return system where

ONE FITTING INSTALLATION—
The "RING" trademark goes between the risers; the Return arrow points in the direction of flow.

TWO FITTING INSTALLATION—
The Monoflo Fitting is used for both Supply and Return. The "RING" trademarks are between the risers; the arrow points in the direction of flow.

Courtesy of ITT Fluid Handling

Fig. 6-15. Monoflow fitting used in hot-water heating systems.

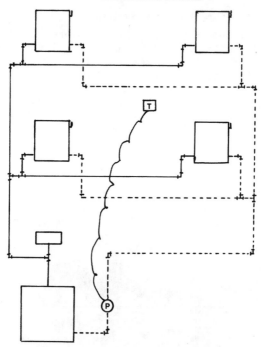

Fig. 6-16. A two-pipe direct-return hot-water heating system.

circulation of that floor or area is controlled by its own pump is
called a zone.

In the reverse-return system shown in Fig. 6-17, the first radiator
supplied from the return main has the *shortest* supply main but has
the *longest* return main. The supply main to each successive radiator
is progressively longer, but as the supply main grows longer the
return main grows progressively shorter. The actual developed

length of supply and return mains to each radiator is approximately equal, therefore the system is balanced, and each radiator should heat evenly. In larger systems with many radiators on each zone, balancing cocks may be needed to ensure an even flow to each

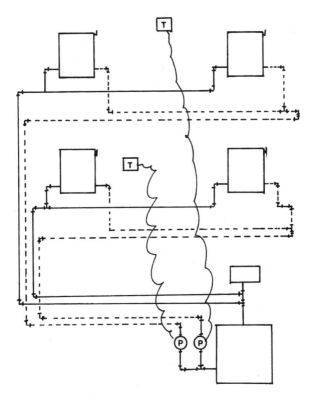

Fig. 6-17. A two-pipe reverse-return hot-water heating system.

individual radiator and to each zone. The initial cost of installing a reverse-return system is higher than for a direct return system, but the advantages of a reverse-return system are well worth the added cost.

TROUBLE-SHOOTING STEAM AND HOT-WATER HEATING SYSTEMS

PROBLEM: Steam does not circulate to end of mains.
Cause: (a) End of main vents not working, (b) in wrong location, (c) boiler dirty; skim off grease, oil, (d) insufficient water-line difference between the low point of the horizontal main and the boiler water line. In vapor systems the ends of dry-return mains should be 24 inches or more above the boiler water line, depending on size of installation and pressure drop. In one-pipe gravity systems this distance should be 18 inches or more.

PROBLEM: Boiler slow to respond.
Cause: (a) May be due to poor draft, (b) boiler too small for load, (c) inferior fuel, (d) improper firing rate, (e) accumulation of clinkers on grate, (f) poor draft.

PROBLEM: Boiler smokes through draft door.
Cause: (a) Air leaks into boiler or breeching, (b) gas outlet from firebox plugged, (c) defective draft; dirty or clogged flues, (d) reduction in breeching size.

PROBLEM: In steam boilers, gauge pressure builds up quickly but steam does not circulate.
Cause: Oil, grease, dirt in boiler; clean boiler.

PROBLEM: Boiler fails to deliver enough heat.
Cause: (a) Poor fuel, (b) poor draft, (c) improper firing

rate, (d) improper piping, (e) boiler too small for load, (f) heating surfaces covered with soot.

PROBLEM: Water line in gauge glass unsteady.
Cause: (a) Grease, oil, dirt in boiler, (b) water column connected to a very active section and therefore not showing actual water level in boiler, (c) boiler operating at excessive output.

PROBLEM: Water disappears from gauge glass.
Cause: (a) Oil, grease, dirt in boiler causing priming, (b) too great a pressure differential between supply and return piping, causing water to back into return, (c) valve closed in return line, (d) water column connected into very active boiler section or thin waterway, (e) improper connections between boilers in battery permitting boiler with excess pressure to push water into boiler with lower pressure, (f) firing rate too high.

PROBLEM: Water is carried into steam main.
Cause: (a) Using boiler beyond rated capacity, (b) grease, oil, dirt in boiler, (c) improper boiler for job, (d) outlet connections too small, (e) firing rate too high, (f) water level in boiler too high.

PROBLEM: Boiler flues soot-up quickly.
Cause: (a) Poor draft, (b) combustion rate too low, (c) excess air in firebox, (d) smoky combustion, (e) improper firing rate.

PROBLEM: Water Hammer
Cause: Water hammer is caused by a shock wave traveling through water. In steam heating systems, the wave is often caused by steam passing at a high velocity over condensate

collected in piping. Steam becomes trapped in pockets of water; rapid condensation causes these slugs of water to collide with considerable force and become audible as water hammer. Water hammer can cause severe damage to a piping system or its components.

Water Hammer in Mains

(a) Water pocket caused by sagging of the main. *Remedy:* Install pipe hangers.

(b) Improper pitch of main. *Remedy:* Check pitch with spirit level; correct where necessary.

(c) Pipe not sized correctly. *Remedy:* Install correct size pipe.

(d) Insufficient water line difference between the low point of the horizontal main and boiler water line. *Remedy:* In one-pipe gravity systems, this distance should be 18 inches or more; in vapor systems, the ends of dry-return mains should be 24 inches or more above waterline depending on installation and pressure drop.

(e) Air valves for venting steam mains in one-pipe gravity or vacuum systems not located properly. *Remedy:* Install air valves in correct locations.

(f) Excessive quantities of water in main due to priming boiler or improper header construction. *Remedy:* All boiler tappings should be used and connected full size to boiler header.

Water Hammer in a Hartford Loop Connection

In a Hartford connection a close nipple should be used between the end of the return and the header drip or equalizing pipe. If the nipple used at this point is too long and the water line of the boiler becomes too low, hammer will result. The remedy is to offset the return piping below the water line of the boiler using a close nipple entering header drip to maintain correct boiler water line. The top of the close nipple should be 2 inches below water line.

Radiator Troubles

Pounding noise in a one-pipe system may be caused by:
    (a) Radiator supply valve partly closed or too small.
    (b) Radiator pitched away from supply valve.
    (c) Vent port of air valve too large, allowing steam to enter radiator too rapidly.

If Radiator does not heat:
    (a) Air in radiator not venting.
    (b) Drainage tongue of air valve damaged or removed.
    (c) Branch supply to radiator too small.
    (d) Vent port of air valve clogged (one-pipe systems).
    (e) Steam pressure higher than maximum working pressure of air valve. (Most likely when steam is supplied through a reducing valve from a high pressure supply.)
    (f) Branch supply improperly pitched causing water pocket.
    (g) Return branch improperly pitched causing water pocket to form and trapping air (in a vapor system).
    (h) In a gas- or oil-fired one-pipe vacuum system, some radiators may not heat if they were not completely heated on a previous firing. Remedy this by changing to a non-vacuum system by using open vents on radiators and mains.

Radiator cools quickly (One-pipe vacuum system):
    (a) Air leakage into system, either through leaky joints or through stuffing box of radiator supply valve if ordinary valve is used.
    (b) Malfunction of vacuum valve preventing formation of vacuum.
    (c) On gas- or oil-fired system, radiator cools quickly due to rapid formation of vacuum. Change to non-vacuum (open) system by changing radiator and main vents to non-vacuum type.

CHAPTER 7

AIR CONDITIONING AND REFRIGERATION

The information in this chapter is designed with three purposes in mind:

1. To explain the basic principles of refrigeration.
2. To acquaint refrigeration servicing personnel with the recent changes brought about by Public Law 101-549. *Public Law 101-549, and its impact on the servicing and repairing of refrigeration and air conditioning equipment, is a very important segment of this chapter.*
3. To provide servicing and trouble-shooting information for air conditioning and refrigeration technicians. But first, let's deal with the basic principles of refrigeration.

Transfer of Heat from One Substance or Element to Another

Thermodynamics is a branch of science that deals with the mechanical action of heat. Refrigeration is a process designed to transfer heat from one medium to another. Heat always travels from a *warm* body to a *colder* body. *Heat exists at any temperature above absolute zero*. Absolute zero, approximately 460°F below zero, is a theoretical term for the lowest temperature possible, the temperature at which *no heat exists*. Heat travels in any one of three ways:

1) Radiation

Radiation is the transfer of heat by waves. Heat from the sun is transferred by radiation.

2) Conduction

Conduction is the transfer of heat through a medium: water or metal. One substance must touch another for heat to be transferred in this way.

3) Convection

Convection is the transfer of heat by means of a fluid medium, either gas or liquid, normally air or water. A hot-water fin tube convector transfers heat from the convector to air.

Heat transfer cannot take place without a *temperature difference*.

Measuring Heat

The British thermal unit (Btu) is the basic unit used to measure heat in the U.S. A Btu is the amount of heat needed to raise the temperature of one pound of water one degree Fahrenheit.

There are several different types of refrigeration and air conditioning equipment, but the basic principles governing air conditioning and refrigeration apply to all systems.

All refrigeration systems depend on two basic principles:

1. A liquid absorbs heat when it boils or evaporates to a gas. Examples:

    A. Freon 12 boils to vapor at –21°F. at atmospheric pressure.
    B. Water boils to a gas (steam) at 212°F. at atmospheric pressure.
    C. Liquid ammonia boils to vapor at –28°F. at atmospheric pressure.

2. As vapor or gas condenses to a liquid form, heat is released.

The boiling point of a liquid is changed by changing the pressure the liquid is under. *Raising* the pressure *raises the boiling point. Lowering* the pressure *lowers the boiling point.*

Fig. 7-1 explains the changes that take place as the refrigerant is circulated through an electrical air conditioning system. We will assume that the thermostat has called for cooling, the compressor is operating, and the fans in the air handling unit and the condenser are energized.

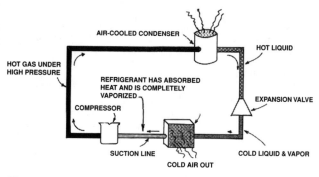

Fig. 7-1. Air conditioning system with an air-cooled condenser.

A typical electric air conditioner operates by introducing a hot liquid under high pressure into an expansion valve. The expansion valve releases the hot liquid, lowering the pressure and causing the hot liquid to expand into a cold liquid plus vapor. The cold liquid plus vapor then passes through the evaporator, absorbing heat from the air surrounding the evaporator. As the liquid+vapor passes through the evaporator coil absorbing heat, it is completely vaporized, changing from a liquid to a cold gas. The compressor pulls the cold gas from the evaporator or cooling coil through the *suction* line as fast as it vaporizes. Next, the compressor compresses the cold gas to a high pressure and, in the process, raises the temperature of the cold gas, changing it to a hot gas. The increased pressure forces the hot gas through the condenser where either air or water removes the heat from the refrigerant gas. The hot gas must be at a higher temperature than the air or water surrounding the condenser in order for heat to flow from the gas to the air or water. As the hot gas gives off heat in the condenser, the gas *condenses* and becomes a hot *liquid,* still under high pressure. The hot liquid enters the expansion valve, and the process begins all over again.

The reason this process works is that the hot gas leaving the compressor is hotter than the air (or water) passing through the condenser. Thus the heat in the gas, absorbed at the evaporator, can flow to the air or water passing through the condenser.

The two types of refrigerating compressors used in air conditioning equipment are reciprocating compressors and centrifugal compressors.

Water is used as a cooling medium for a condenser in many large air conditioning systems. Cooled water from a cooling tower basin is pumped through the water jacket of a condenser. The cooled water absorbs heat from the hot gas in the refrigerant piping coils in the condenser. The water, now hot from the absorbed heat, is circulated to the cooling tower. There it runs into a distributing trough and drips down through the baffles in the tower. As the water drips down, it gives off the absorbed heat. The cooled water is now ready to be pumped back through the condenser to again absorb heat. Some water is lost through evaporation, so a make-up water valve is installed in the basin to maintain the proper water level. A typical cooling tower is shown in Fig. 7-2. A piping diagram of a system using a water-cooled condenser and a cooling tower is shown in Fig. 7-3.

Gauge Manifold

Whenever an air conditioning unit must be serviced, the first thing the serviceman does is to connect a set of gauges to the unit. The most important tool used by the serviceman is the gauge set. It is used for checking pressures, adding refrigerant, and many other purposes. Space available will not permit instructions for the use of this tool, nor is it necessary, because the serviceman for whom the balance of this chapter is written knows how to use it.

Public Law 101-549, The Clean Air Act

Title VI of the Clean Air Act Amendments of 1990, "Stratospheric Ozone Protection," is a comprehensive program to restrict the production and use of chemicals that reduce the amount

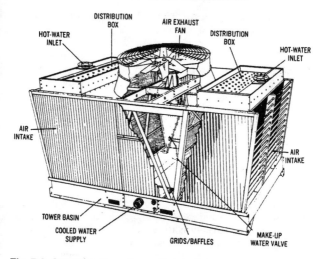

Fig. 7-2. Cutaway view of a cooling tower.

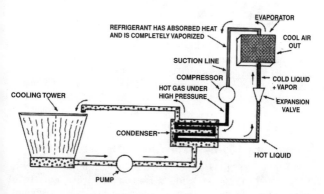

Fig. 7-3. An air conditioning system with a cooling tower and a water-cooled condenser.

of ozone in the stratosphere. Title VI replaces Part B of Title I of the Clean Air Act of 1977.

This legislation is designed to bring about changes in the way CFCs (Chlorofluorocarbons) and HCFCs (Hydrochlorofluorocarbons) are handled and disposed of. The legislation was enacted because a seasonal "hole in the sky," a massive drop in stratospheric ozone levels was discovered in the Southern Hemisphere.

The Clean Air Act establishes a program that includes production targets and use limits and bans, recycling requirements, venting prohibitions, labeling requirements, and safe alternative measurements.

Public Law 101-549, the Clean Air Act Amendments, is now in effect, and it is drastically changing the way refrigeration and air conditioning equipment is serviced and installed. Installation and servicing of refrigeration and air conditioning equipment provide employment for pipe fitters specializing in these fields. The intent of P.L. 101-549 is to prevent, so far as possible, the venting of ozone-depleting substances into the atmosphere. Section 608, subsection (c), paragraph 1, reads: "It is unlawful for any person in the course of maintaining, servicing, repairing, or disposing of an appliance or industrial process refrigeration to knowingly vent or otherwise knowingly release or dispose of Class I or Class II refrigerants in a manner that permits such substances to enter the environment." The penalty for intentionally allowing the venting of ozone-depleting chemicals, CFCs (chlorofluorocarbons) and HCFCs (hydrochlorofuorocarbons), can be severe. On December 2, 1992, the federal Environmental Protection Agency (EPA) fined a New England contractor $18,101 for venting an HCFC refrigerant.

Contractors (*and employees*) take note: If a contractor sends a technician on a service call that requires opening up a charged system and he or she does not have a *certified* refrigerant recovery or recycling unit on the job, he *and you* are technically in violation of the Clean Air Act. Failure to have the recovery or recycle unit on the job represents a strong presumption of guilt.

Three Rs are the key to understanding the purpose of Public Law 101-549, *Recover, Recycle, and Reclaim.*

To *recover* refrigerant means to remove refrigerant in any condition from an appliance without necessarily testing or reprocessing it in any way.

Since July 1, 1992, CFC being removed from systems *must be recovered.* Recovery is not a choice; *it is the law.* It is very important to record the weight of the CFC removed. This will be taken into account when the unit is charged after repairs are made.

To *recycle* refrigerant means to extract refrigerant from an appliance and clean refrigerant for re-use without meeting all the requirements for reclamation. Recycled refrigerant is refrigerant that is cleansed using oil separation and single or multiple passes through devices such as replaceable core filter-driers which reduce moisture, acidity, and particulate matter. Recycling is usually done at the job site.

To *reclaim* refrigerant means to reprocess refrigerant to at least the purity specified in the ARI Standard 700-1988, Specifications for Fluorocarbon Refrigerants (Appendix A to 40 CFR part 82, subpart F), and to verify this purity using the analytical methodology prescribed in the ARI Standard 700-1988.

In general, reclamation involves the use of processes or procedures *available only* at a reprocessing or manufacturing facility. Contaminated refrigerants and/or oils must be recovered and stored in containers, returned to a collection point (it may be the supply house the customer deals with), and returned to the reprocessor or manufacturer.

Most CFC based systems in use today use a mineral oil lubricant. Studies have shown that there is poor miscibility between many of the alternatives to CFCs and mineral oil. New oils have been developed that are more compatible with alternative refrigerants.

Now we come to the problem of selecting the equipment needed to recover and/or recycle the refrigerants. Since it is common practice to install refrigeration components on rooftops or other relatively inaccessible locations, both size and weight of recovery and/or recycling units must be considered. As the full impact of Public Law 101-549 on refrigeration and air conditioning contractors is felt, the demand for recovering and recycling units may, for a time, exceed the available supply. Table 7-1 shows a listing of Class I and Class II substances used in refrigeration and air conditioning systems.

A lightweight portable recovery unit for small system services is shown in Fig. 7-4.

Courtesy Robinaire Div., SPX Corp.

Fig. 7-4. A lightweight refrigerant recovery unit.

The unit shown in Fig. 7-5 is designed with on-site recycling capability enabling the operator to recharge the system with refrigerant substantially cleaner than when it was recovered from the system.

A recovery/recycling unit with an oil separator/heat exchanger is shown in Fig. 7-6. This unit is designed to recover/recycle large quantities of refrigerant in high ambient temperatures.

Fig. 7-5. A recovery/recycling unit.

Used and/or contaminated refrigerant gases are returned to service centers which return the contaminated products to the manufacturer for reclaiming and reuse. The refrigerant supply houses that the contractor deals with can supply information on recovery/recycling equipment and refrigerant/oil disposal.

The scope of work for pipe fitters in the air conditioning and refrigeration fields includes the unloading, setting, installation, and repairing of chillers, compressors, condensers, pumps, and, if used, pneumatic control systems. The installation and servicing of pneumatic control systems is another specialized branch of the pipe fitter's trade. Various control companies operate schools teaching the installation and servicing of pneumatic control systems.

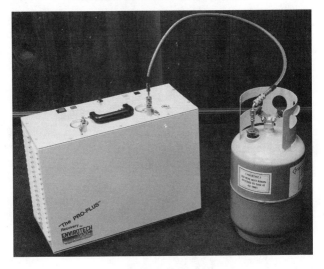

Courtesy Envirotech Systems

Fig. 7-6. A recovery/recycling unit with tank.

The following servicing and troubling-shooting instructions will help in diagnosing problems with refrigeration and air conditioning systems.

Refrigerant Pressures

Twelve of the most common refrigerants are listed in Table 7-2. Each refrigerant is listed in a separate column, and the pressures are listed in *pounds per square in. gauge (psig)* or in *inches of mercury* if the pressures are below zero pound gauge pressure. The refrigeration serviceman's bourdon tube gauge reads 0 psig (lbs. sq. in. gauge) when not connected to a pressure-producing source. Pressures existing below 0 psig are actually negative readings on the gauge and are referred to as inches of vacuum. The vacuum gauge is calibrated in the equivalent of inches of mercury. The column at the left of Table 7-2 shows the temperature in degrees F. The temperatures are given in five (5) degree increments. The extremely low temperatures are seldom encountered. The valves or controls can be adjusted to achieve a temperature falling between the five degree increments. In order to determine the vaporization pressure in the evaporator, the temperature of the liquid must be determined.

Example: With ammonia as the refrigerant, we want the ammonia to boil at 10° above zero F. The refrigerant plant will cool brine. The brine, cooled to approximately 25°, will be pumped to the cooling coils. Checking the temperature column at the 10° line, then following the line to the right under the heading AMMONIA, the pressure should be 23.8 pounds. The head of condensing pressures can then be checked. This is done by taking the temperature of the outlet water if the refrigerating unit is water-cooled. When we assume that the water is leaving the condenser at a temperature of 90°F, a reference to Table 7-2 indicates that the head pressure should be approximately 166 pounds.

If air-cooled apparatus is encountered, the temperature of the air passing over the condenser must first be determined. Let us assume that the air passing over the condenser is at a temperature of 70°F. Observe in Table 7-3, that one temperature column gives the temperature of the liquid in the evaporator. In our particular case, the ammonia is evaporating at a temperature of 10°F. Therefore find this specific temperature in the left hand column and continue horizontally to that point in the column headed 70°F. At this junction the figure "35" is shown. This figure "35" is added to the air temperature, 70°F., giving a total of 105°F. Turn to Table 7-2 and find the 105° line, corresponding to the temperature existing in the condenser. Table 7-2 gives the temperature at which boiling or vaporization of each particular refrigerant takes place; the pressure at which this occurs is listed under the refrigerant type.

If the refrigerant is ammonia at 15.7 psig, the liquid will boil at 0°F. If you desire a colder liquid, alter the pressure. This is accomplished in the refrigerating unit by adjusting the temperature control, if a thermostat is the controller (as with a low-side float), or by adjusting the thermostat or pressurestat in either a high-side or low-side float system. The continued operation will lower the evaporator temperature and the temperature of the room in which the evaporator is located. With an expansion valve system, if a thermostat is employed to start and stop the system, the thermostat can be adjusted for a higher or lower temperature. Before and after the adjustment, the frost line on the evaporator should be checked.

If the expansion valve is opened to admit more refrigerant so that the pressure is raised and its boiling temperature is increased, the suction line may frost back. Also, if the expansion valve is closed so that less refrigerant enters the evaporator, the compressor will be unable to hold at a lower pressure and a lower temperature. When this is done, only a portion of the coil may be effective. A coil must contain liquid refrigerant to be able to refrigerate. Therefore, if only a part of the coil is frosted, the other part might as well be taken out since it is doing no work. It is best to maintain the highest pressure possible so that the compressor will work on the densest vapor without having the suction line frost back.

It may be desirable, however, to make a change in the suction pressure to alter the temperature of the coils. If a coil temperature at 20°F. is desired with ammonia, the expansion valve must be adjusted to maintain a constant pressure of 33.5 psig in the evaporator. If a 40° temperature is desired, a 58.6 psig reading on the low side would have to be maintained. If a 40° temperature were required in the evaporator of a methyl-chloride system, the expansion valve would have to be adjusted to hold a constant pressure of 28.1 psig.

Table 7-1. Class I and Class 2 Substances

SEC. 602. LISTING OF CLASS I AND CLASS II SUBSTANCES.

(a) LIST OF CLASS I SUBSTANCES.—Within 60 days after enactment of the Clean Air Act Amendments of 1990, the Administrator shall publish an initial list of class I substances, which list shall contain the following substances:

Group I
  chlorofluorocarbon-11 (CFC-11)
  chlorofluorocarbon-12 (CFC-12)
  chlorofluorocarbon-113 (CFC-113)
  chlorofluorocarbon-114 (CFC-114)
  chlorofluorocarbon-115 (CFC-115)

Group II
  halon-1211
  halon-1301
  halon-2402

Group III
  chlorofluorocarbon-13 (CFC-13)
  chlorofluorocarbon-111 (CFC-111)
  chlorofluorocarbon-112 (CFC-112)

(b) LIST OF CLASS II SUBSTANCES.—Simultaneously with publication of the initial list of class I substances, the Administrator shall publish an initial list of class II substances, which shall contain the following substances:

  hydrochlorofluorocarbon-21 (HCFC-21)
  hydrochlorofluorocarbon-22 (HCFC-22)
  hydrochlorofluorocarbon-31 (HCFC-31)
  hydrochlorofluorocarbon-121 (HCFC-121)
  hydrochlorofluorocarbon-122 (HCFC-122)
  hydrochlorofluorocarbon-123 (HCFC-123)
  hydrochlorofluorocarbon-124 (HCFC-124)
  hydrochlorofluorocarbon-131 (HCFC-131)
  hydrochlorofluorocarbon-132 (HCFC-132)
  hydrochlorofluorocarbon-133 (HCFC-133)
  hydrochlorofluorocarbon-141 (HCFC-141)
  hydrochlorofluorocarbon-142 (HCFC-142)
  hydrochlorofluorocarbon-221 (HCFC-221)

Table 7-1. Class I and Class 2 Substances *(Cont'd)*

SEC. 602. LISTING OF CLASS I AND CLASS II SUBSTANCES.

hydrochlorofluorocarbon-222 (HCFC-222)
hydrochlorofluorocarbon-223 (HCFC-223)
hydrochlorofluorocarbon-224 (HCFC-224)
hydrochlorofluorocarbon-225 (HCFC-225)
hydrochlorofluorocarbon-226 (HCFC-226)
hydrochlorofluorocarbon-231 (HCFC-231)
hydrochlorofluorocarbon-232 (HCFC-232)
hydrochlorofluorocarbon-233 (HCFC-233)
hydrochlorofluorocarbon-234 (HCFC-234)
hydrochlorofluorocarbon-235 (HCFC-235)
hydrochlorofluorocarbon-241 (HCFC-241)
hydrochlorofluorocarbon-242 (HCFC-242)
hydrochlorofluorocarbon-243 (HCFC-243)
hydrochlorofluorocarbon-244 (HCFC-244)
hydrochlorofluorocarbon-251 (HCFC-251)
hydrochlorofluorocarbon-252 (HCFC-252)
hydrochlorofluorocarbon-253 (HCFC-253)
hydrochlorofluorocarbon-261 (HCFC-261)
hydrochlorofluorocarbon-262 (HCFC-262)
hydrochlorofluorocarbon-271 (HCFC-271)

The initial list under this subsection shall also include the isomers of the substances listed above. Pursuant to subsection (c), the Administrator shall add to the list of class II substances any other substance that the Administrator finds is known or may reasonably be anticipated to cause or contribute to harmful effects on the stratospheric ozone layer.

Table 7-2 lists the temperature of the liquid refrigerant boiling at the constant pressure given. One factor can not be varied without varying the other. The only way to obtain another boiling temperature is to adjust the expansion valve to maintain the pressure that corresponds to the temperature desired. If the liquid is contained in a vessel or cylinder and the vessel is closed (such as in a supply drum), you will observe that the pressure in such a vessel is from one to five pounds greater than the temperature of the drum. This is due to the fact that pressure has built up and prohibits further evaporation. If a system contains an unknown refrigerant, such as one without an odor, the kind may be determined by taking a pressure reading of the refrigerant at rest, that is, while the unit is not operating. The pressure that corresponds nearest to the reading will identify the particular refrigerant being tested.

Table 7-2 is based on determining, from a pressure reading of the low side or evaporator, just what the temperature of the boiling liquid may be. With a methyl-chloride system, if the low pressure gauge reads 2 psig, the temperature corresponding to this pressure is given as −5°F. in Table 7-2.

With the temperature of the evaporator known (−5°F.) refer to Table 7-3. The first column in Table 7-3 on the left hand side is the evaporator temperature. The other columns refer to the temperature of the water or air passing over the condenser. Assume that the room temperature is 70°F. and that the air is passing over the condenser at this temperature. Since the evaporator temperature is known, find this temperature in the first column and then find the proper column for the coolant temperature (70°F.). At the intersection under these conditions, the figure 20 is given. This is added to the initial temperature, thus: 20° + 70° = 90°. Take this figure (90°) and refer to Table 7-2. Run down the temperature column to 90° and then across to the methyl-chloride column. You will find that a head or condensing pressure of about 87.3 pounds can be expected. If the actual pressure reading varies from this, check the condition of the apparatus.

Table 7-2. Pressure in Pounds Per Square In. (Gauge) or Inches of Vacuum Corresponding to Temperature in Degrees F. for Various Common Refrigerants.

Tem., °F.	Ammonia	Sulfur dioxide	Methyl chloride	Ethane	Propane	Ethyl chloride	Carbon dioxide	Butane	Isobutane	Freon F-12	Correne	Methyl formate
-40	8.7"	23.5"	15.7"	99.8#	1.5#		131.1#			11.0"		
-35	5.4"	22.4"	14.4"	109.8#	3.4#		156.3#			8.4"		
-30	1.6"	21.1"	11.6"	120.3#	5.6#		163.1#			5.5"		
-25	1.3#	19.6"	9.2"	132.0#	8.0#		176.3#			2.3"		
-20	3.6#	17.9"	6.1"	144.8#	10.7#	25.3"	205.8#		14.6"	0.5#		
-15	6.2#	16.1"	2.3"	157#	13.6#	24.5"	225.8#		13.0"	2.4#		
-10	9.0#	13.9"	0.2"	172#	16.7#	23.6"	247.0#		11.0"	4.5#	28.1"	
-5	12.2#	11.5"	2.0#	187#	20.0#	22.6"	269.7"		8.8"	6.8#	27.8"	
0	15.7#	8.8"	3.8#	204#	23.5#	21.5"	293.9#	15.0"	6.3"	9.2#	27.5"	26.5"
+5	19.6#	5.8"	6.2#	221#	27.4#	20.3"	319.9#	12.2"	3.3"	11.9#	27.1"	25.9"
+10	23.8#	2.6"	8.6#	239#	31.4#	18.9"	347.1#	11.1"	0.2"	14.7#	26.7"	25.4"
+15	28.4#	0.5#	11.2#	257#	35.9#	17.4"	376.3#	8.8"	1.6#	17.7#	26.2"	24.7"
+20	33.5#	2.4#	13.6#	227#	40.8#	15.8"	407.3#	6.3"	3.5#	21.1#	25.6"	24.0"
+25	39.0#	4.6#	17.2#	292#	46.2#	14.0"	440.1#	3.6"	5.5#	24.6#	24.9"	23.1"
+30	45.0#	7.0#	20.3#	320#	51.6#	12.2"	474.9#	0.6"	7.6#	28.5#	24.3"	22.3"
+35	51.6#	9.6#	24.0#	343#	57.3#	10.1"	511.7#	1.3#	9.9#	32.6#	23.5"	21.1"

Note " inches of mercury
Note # PSIG

continues

Table 7-2. Twelve Common Refrigerants

Pressure in Pounds Per Square In. (Gauge) or Inches of Vacuum Corresponding to Temperature in Degrees F. for Various Common Refrigerants. (Cont.)

Tem., °F.	Ammonia	Sulfur dioxide	Methyl chloride	Ethane	Propane	Ethyl chloride	Carbon dioxide	Butane	Isobutane	Freon F-12	Correne	Methyl formate
+40	58.6#	12.4#	28.1#	368#	63.3#	8.0"	550.7#	3.0#	12.2#	37.0#	22.6"	20.0"
+45	66.3#	15.5#	32.2#	390#	69.9#	5.4"	591.8#	4.9#	14.8#	41.7#	21.7"	18.7"
+50	74.5#	18.8#	36.3#	413#	77.1#	2.3"	635.3#	6.9#	17.8#	46.7#	20.7"	17.3"
+55	83.4#	22.4#	41.7#	438#	84.6#	0.3"	681.2#	9.1#	20.8#	52.0#	19.5"	15.7"
+60	92.9#	26.2#	46.3#	466#	92.4#	1.9#	729.5#	11.6#	24.0#	57.7#	18.2"	14.0"
+65	103.1#	30.4#	53.6#	496#	100.7#	3.3#	780.4#	14.2#	27.5#	63.7#	16.7"	11.9"
+70	114.1#	34.9#	57.8#	528#	109.3#	6.2#	834.0#	16.9#	31.1#	70.1#	15.1"	9.8"
+75	125.8#	39.8#	64.4#	569#	118.5#	8.3#	890.4#	19.8#	35.0#	76.9#	13.4"	7.3"
+80	138.1#	45.0#	72.3#	610#	128.1#	10.5#	949.6#	22.9#	39.2#	84.1#	11.5"	4.9"
+85	151.7#	50.9#	79.4#	657#	138.4#	12.9#	1011.3#	26.2#	43.9#	91.7#	8.4"	2.4"
+90	165.9#	56.5#	87.3#	693#	149#	15.4#		29.8#	48.6#	99.6#	7.3"	0.5#
+95	181.1#	62.9#	95.6#		160#	18.2#		33.2#	53.7#	108.1#	5.0"	2.1#
+100	197.2#	69.8#	102.3#		172#	21.0#		37.5#	59.0#	116.9#	2.4"	3.8#
+105	214.2#	77.1#	113.4#		185#	24.3#		41.7#	64.6#	126.2#	0.19#	5.8#
+110	232.2#	85.1#	118.3#		197#	27.3#		46.1#	70.4#	136.0#	1.6#	7.7#
+115	251.5#	93.5#	128.6#		207.6#	31.6#			76.7#	146.5#	3.1#	10.4#
+120	271.7#	106.4#	139.3#		218.3#	35.5#			84.3#	157.1#	4.7#	13.1#
+125	293.1#	111.9#	150.3#		232.3#	39.5#			90.1#	168.6#	6.6#	15.7#
+130	315.6#	121.9#	161.3#		246.3#	44.0#			97.3#	180.2#	8.4#	18.2#

Table 7-3. Factors to Be Added to Initial Temperatures of Coolant to Determine Condenser Temperatures

Evap.	Initial Coolant Temp. °F				
°F	60°	70°	80°	90°	100°
−30	15	15	15	10	10
−25	15	15	15	15	10
−20	20	20	15	15	15
−15	20	20	20	15	15
−10	20	20	20	15	15
−5	20	20	20	15	15
0	25	25	20	20	15
+5	30	30	30	25	20
+10	35	35	30	25	20
+15	40	35	30	25	20
+20	40	35	30	25	25
+25	45	40	35	30	30
+30	50	45	40	35	35
+35	50	50	45	45	40

It is essential to obtain a pressure reading on the low or evaporator side so that the temperature of the refrigerant can be determined. If the reading indicates too high a temperature, the expansion valve will have to be adjusted to maintain the proper pressure and, of course, the proper evaporating temperature. Head pressures are important; excessive pressures indicate something out of the ordinary.

Much of the equipment required for refrigeration and air conditioning is large and heavy, and many times this equipment must be placed on roofs or in locations where a crane must be used to hoist the equipment. The crane operator in most cases can not see beyond the edge of a roof or inside an enclosure, and signals must be used to guide the operator while hoisting and setting equipment. The signals shown in Appendix Table A-7 are the standard universally used ANSI/ASME signals for directing a crane operator.

The information on the following pages will aid in servicing refrigeration and air conditioning equipment.

AIR CONDITIONING SYSTEM TROUBLE CHART

Compressor Will Not Start

Possible Cause	Possible Remedy
Thermostat setting too high.	Reset thermostat below room temperature.
High head pressure.	Reset starter overload and determine cause of high head pressure.
Defective pressure switch.	Repair or replace pressure switch.
Loss of refrigerant charge.	Check system for leaks.
Compressor frozen.	Replace compressor.

Compressor Short Cycles

Possible Cause	Possible Remedy
Defective thermostat.	Replace thermostat.
Incorrect setting on low side of pressure switch.	Reset low-pressure switch differential.
Low refrigerant charge.	Check system for leaks; repair and add refrigerant.
Defective overload.	Replace overload.
Dirty or iced evaporator.	Clean or defrost evaporator.
Evaporator blower and motor belts slipping.	Tighten or replace belts.
Dirty or plugged filters.	Clean or replace air filters.

Compressor and Condenser Fan Motor Will Not Start

Possible Cause	Possible Remedy
Power failure.	Check electrical wiring back to fuse box.
Fuse blown.	Replace blown or defective fuse.
Thermostat setting too high.	Reduce temperature setting of room thermostat.
Defective thermostat.	Replace or repair thermostat.
Faulty wiring.	Check wiring and make necessary repairs.

Defective controls.	Check and replace defective controls.
Low voltage.	Reset and check for cause of tripping.
Defective dual-pressure control.	Replace the control.

Compressor Will Not Start, But the Condenser Fan Motor Runs

Possible Cause	Possible Remedy
Faulty wiring to compressor.	Check compressor wiring and repair.
Defective compressor motor.	Replace the compressor.
Defective compressor overload (single phase only).	Replace overload.
Defective starting capacitor (single phase only).	Replace capacitor.

Condenser Fan Motor Will Not Start, But Compressor Runs

Possible Cause	Possible Remedy
Faulty wiring to fan motor.	Check fan motor wiring and repair.
Defective fan motor.	Replace fan motor.

Condenser Fan Motor Runs, But the Compressor Hums and Will Not Start

Possible Cause	Possible Remedy
Low Voltage.	Check line voltage. Determine the location of the voltage drop.
Faulty wiring.	Check wiring and make necessary repairs.
Defective compressor.	Replace compressor.
High head pressure.	Check head pressure and complete operation of system to remove the cause of the high pressure condition.
Failure of one phase (three-phase units only).	Check fuses and wiring.
Defective start capacitor (single phase only).	Replace capacitor.
Defective potential relay (single phase only).	Replace relay.

Compressor Starts, But Cycles on Overload

Possible Cause	Possible Remedy
Low voltage.	Check line voltage. Determine the location of the voltage drop.
Faulty wiring.	Check wiring and make necessary repairs.
Defective running capacitor (single phase only).	Replace capacitor.
Defective overload.	Replace overload.
Unbalanced line (three-phase only).	Check wiring: call power company.

Evaporator Fan Motor Will Not Start

Possible Cause	Possible Remedy
Power failure.	Check electrical wiring back to fuse box.

Compressor Runs Continuously

Possible Cause	Possible Remedy
Excessive load.	Check for excessive outside air infiltration and excessive source of moisture.
Air or noncondensable gases in the system.	Purge System.
Dirty condenser.	Clean condenser.
Condenser blower and motor belts slipping.	Tighten or replace belts.
Thermostat setting too low.	Reset thermostat.
Low refrigerant charge.	Check system for leaks; repair and add refrigerant.
Overcharge of refrigerant.	Purge and remove excess refrigerant.
Compressor valves leaking.	Replace compressor.
Expansion valve or strainer plugged.	Clean expansion valve or strainer.

System Short of Capacity

Possible Cause	Possible Remedy
Low refrigerant charge.	Check system for leaks; repair and add refrigerant.

System Short of Capacity *(Cont.)*

Incorrect superheat setting of expansion valve.	Adjust superheat to 10°F.
Defective expansion valve.	Repair or replace valve.
Air or noncondensable gases in the system.	Purge system.
Condenser blower and motor belts slipping.	Tighten or replace belts.
Overcharge of refrigerant.	Purge excess refrigerant.
Compressor valves leaking.	Replace compressor valves.
Expansion valve or strainer plugged.	Clean valve or strainer.
Condenser air short-circuiting.	Remove obstructions or causes of short.

Head Pressure Too High

Possible Cause	Possible Remedy
Overcharge of refrigerant.	Purge excess refrigerant.
Air or noncondensable gases in system.	Purge system.
Dirty condenser.	Clean condenser.
Condenser blower and motor belts slipping.	Tighten or replace belts.
Condenser air short-circuiting.	Remove obstructions or causes of short-circuiting air.

Compressor and Condenser Fan Motor Will Not Start

Possible Cause	Possible Remedy
Power failure.	Check electrical wiring back to fuse box.
Fuse blown.	Replace blown or defective fuse.
Thermostat setting too high.	Reduce temperature setting of room thermostat.
Defective thermostat.	Replace or repair thermostat.
Faulty wiring.	Check wiring and make necessary repairs.
Defective controls.	Check and replace defective controls.
Low voltage.	Reset and check for cause of tripping.
Defective dual-pressure control.	Replace the control.

Head Pressure Too Low

Possible Cause	Possible Remedy
Low refrigerant charge.	Check system for leaks: repair and add refrigerant.
Compressor valves leaking.	Replace compressor valves.

Suction Pressure Too High

Possible Cause	Possible Remedy
Excessive load on system.	Remove conditions causing excessive load.
Expansion valve is stuck in "Open" position.	Repair or replace expansion valve.
Incorrect superheat setting of expansion valve.	Adjust superheat setting to 10°F.

Suction Pressure Too Low

Possible Cause	Possible Remedy
Low refrigerant charge.	Check system for leaks: repair and add refrigerant.
Expansion valve or strainer plugged.	Clean expansion valve or strainer.
Incorrect superheat setting of expansion valve.	Adjust superheat setting to 10°.
Evaporator air volume low.	Increase air over evaporator.
Stratification of cool air in conditioned area.	Increase air velocity through grilles.

Compressor Is Noisy

Possible Cause	Possible Remedy
Worn or scored compressor bearings.	Replace compressor.
Expansion valve is stuck in "Open" position or defective.	Repair or replace expansion valve.
Overcharge of refrigerant or air in system.	Purge system.
Liquid refrigerant flooding back to compressor.	Repair or replace expansion valve.

Compressor Is Noisy *(Cont.)*

Possible Cause	**Possible Remedy**
Shipping or hold-down bolts not loosened.	Loosen compressor hold-down bolts so compressor is freely floating in mountings.

SERVICING

A malfunctioning system may be caused by one part of the system or a combination of several parts. For this reason, it is necessary and advisable to check the more obvious causes first. Each part of the system has a definite function to perform, and if this part does not operate properly, the performance of the entire air conditioning system will be affected.

To simplify servicing, especially a system that is unfamiliar, the serviceman should remember that the refrigerant, under proper operating conditions, travels through the system in one specified direction. Remembering this, the path of the refrigerant can be traced through any refrigerating system. For instance, beginning at the receiver, the refrigerant will pass through the liquid shutoff valve. If this valve is partially closed, or is completely closed or clogged, no refrigeration is possible, even though the unit itself can and will operate.

Strainers and Filters

Most refrigerating units have a strainer or filter in the liquid line. If this device becomes clogged or filled with dirt, the liquid refrigerant will be unable to pass. Filters are usually designed to hold a certain amount of dirt, scale, metallic particles, etc., and still function, but sometimes through carelessness, an excessive amount of extraneous matter may be allowed to enter the system during assembly or servicing. The remedy is obvious—the dirt and foreign matter must be removed to permit the refrigerant to pass through the filter.

Copper Tubing

Many smaller air conditioning units make use of copper tubing for the liquid and suction lines. Since copper is not structurally as strong as iron or steel, it is relatively easy to collapse the copper line by an accidental blow, thus preventing the circulation of the refrigerant. When this happens, a new line or section of the line must be installed.

Expansion Valves

Expansion valves must be properly adjusted and in perfect working order to maintain the correct pressure in the low side of the line. In many cases, oil from the compressor crankcase enters and remains in the evaporator, occupying the space intended for the refrigerant. Naturally, this reduces the refrigerating effect and service problems develop. The oil must be drained from the evaporator coil, and the compressor crankcase inspected to make sure the oil level is correct.

Compressors

Compressors will start developing trouble after long use as the pistons and rings, as well as the other components, become worn. This wear allows some of the high-pressure gas to leak by the pistons and rings, so that proper compression cannot be obtained, and the compressor becomes inefficient and unable to take care of the load. New rings, pistons, and connecting rods usually are all that are required to bring the compressor back to its proper performance. Suction and discharge valves may stick, crack, or fail entirely. Sometimes, a slight warpage or piece of dirt under them results in improper operation. Seals around the crankshaft may leak and require repacking or replacement.

Condensers

The condensers may become dirty and inefficient, thereby resulting in high pressure, loss of efficiency, and an increase in the power requirement.

Ducts

Some air conditioning ducts are provided with insulation on the inside which may tear loose and flap, resulting in noise.

Water Jets

Water jets or spray heads (if used) may become clogged or worn. Automatic water valves may get out of adjustment, and the flow of water may become restricted because of clogging of the water strainer.

SERVICING CHECKS

The proper diagnosis of air conditioning system service problems can only be determined by intelligent thought, patience, and diligence. Each and every contributory factor must be considered and eliminated before going to some other cause. The various important steps in checking a system are listed as follows:

1. Determine the refrigerant used in the system. This is an important factor since each refrigerant has its own operating characteristics, such as pressure-temperature differences.
2. Install a gauge test set. If the unit is a large one, it will have a set of gauges as part of the equipment. The gauge test will indicate the condition of the refrigerant by checking the pressure-temperature relations. Put a thermometer on the evaporator coil, near the expansion valve, and obtain a reading. This temperature reading, along with a back-pressure reading, will indicate the refrigerant condition. Then check for improper expansion-coil pressure.
3. A low-pressure reading may be caused by a shortage of refrigerant, the presence of ice or water in the adjustment side of the expansion valve, moisture in the refrigerant system, plugged screens, partly closed liquid shutoff or service valves, excessive charge, or air in the system.

Improper valve settings, blocked air circulation over the condenser, cooling air passing over the condenser at too high a temperature, reversed rotation of the motor, bent fan blades, or a clogged condenser are indicated by high head pressures.

4. Frost or a sweat line on the coil should be noted. A coil is only active up to its frost or sweat line. All lengths of tubing beyond the frost or sweat line are inactive, since they do not receive liquid refrigerant. For greatest efficiency, the entire coil must have frost or sweat, and if the liquid line is obstructed, the screen plugged, or the expansion improperly set, the proper amount of liquid refrigerant will not enter the evaporator, and only a limited portion of the coil will frost or sweat, depending on the type of liquid refrigerant used.

5. An increase or overabundance of refrigerant will cause the entire coil, and possibly the suction line, to frost or sweat. This may also be caused by a leaky needle, improper adjustment, ice in the adjustment side of the expansion valve, or fused thermostat contacts. Most complaints with regard to excessive or insufficient frosting are due to weather conditions.

6. Improper operation will occur if refrigerant has escaped from the system. The first thing to do when leakage has occurred is to detect and repair the leak. Indications of refrigerant shortage include hissing at the expansion valve, a warm or hot liquid line, little or no frost on the expansion valve or coil, continuous operation, low head pressure, and bubbles in the sight glass if it is inserted into the liquid line and used to test for refrigerant shortage.

7. Bubbles may form in the sight glass if the head pressure is under 120 pounds with F-12 and under 100 pounds with methyl chloride. An excellent indication of leakage is the presence of oil on a joint or fitting; methyl chloride and F-12 dissolve oils, and when a leak occurs, the

escaped refrigerant evaporates in the atmosphere, leaving the oil behind. Refrigerant must be added to system having a refrigerant shortage until the bubbles in the sight glass cease, or until the hissing sound at the expansion valve is eliminated.

8. Check for improper installation. Compressor units and low sides must be level. Tubes and fittings forming the liquid and suction lines must be the proper size. Baffles, ducts, and eliminators must be properly located and not obstructed. Ducts must be insulated properly, or short circuiting of air currents may take place, causing the ducts to sweat. Liquid and suction lines must be checked for pinches, sharp or flattened bends, and obstructions. Lines should not be run along the ceiling in a hot room, and the lines should not run adjacent to any active hot-water or steam pipes.

9. The location and installation of the thermostat must be checked. The thermostatic-switch bulb should be installed in a location where average temperature conditions exist. The thermostatic bulb, or switch itself (if self contained), should not be placed where an inrush of warm air, such as that caused by the opening of doors or windows, will cause the mechanism to cut in prematurely. Use thermostatic control switches with a minimum of tubing—gas will condense in long runs of tubing and the condensate will produce erratic operation. If a thermostatic bulb is used to control a liquid bath or brine, the bulb should be housed in a dry well or cavity, which should be located where average temperature conditions will be stable.

10. The condition of the thermostatic bulb should be checked. Apply heat gently by means of a cloth saturated with hot water or else hold the bulb in the hand. If the contacts do not close after applying heat, the thermostatic element is discharged. Dirty or oxidized contacts can also cause defective operation.

11. Starter fuses should be checked and, if blown, replaced with fuses of the proper size. Determine the cause of the trouble—shortage of oil, air in the system, overcharge, misalignment, high back pressure, lack of air or water over the condenser or motor—any of which may cause the fuses to blow. If the trouble is not corrected, serious damage can result in a relatively short time.

12. Filters and/or screens must be checked. A clogged screen or filter, or a pinched or clogged liquid line, will produce the same trouble as a leaky or stuck expansion valve, depending on the degree of the obstruction.

13. A leaky or stuck expansion-valve needle will be indicated by a low pressure in the evaporator side and continuous operation of the unit. Sometimes a low-pressure control is wired in series with the thermostatic control and motor. When this is done, the low-pressure control will cut out, and the unit will remain inoperative.

COMMON TROUBLES IN AUTOMATIC EXPANSION VALVE SYSTEMS WITH THERMOSTATIC CONTROL

Refrigerant Shortage

1. Warm or hot liquid line.
2. Hissing sound at expansion valve.
3. Low head or condensing pressure.
4. Evaporator not entirely chilled.
5. Poor refrigeration.
6. Low-side pressure may be high if only gas is entering the evaporator.

Poor Refrigeration

1. Heavy coat of frost or ice on evaporator.
2. Refrigerant shortage.

3. Thermostat not adjusted properly.
4. Thermostat defective.
5. Thermostat shifted and not level.
6. Thermostat shielded by covering.
7. Thermostat in draft.
8. Stuck expansion valve.
9. Expansion valve set too low; allows only a portion of the evaporator to be effective.
10. Liquid line pinched or restricted.
11. Suction line pinched or restricted.
12. Compressor valves defective, broken, or sticking.
13. Strainer on liquid or suction line clogged.
14. Partially closed liquid- or suction-line valves.
15. Ice or moisture in adjustment side of expansion valve.
16. Ice freezing in seat of expansion valve.
17. High head pressure.
18. Compressor losing efficiency through wear.
19. Compressor may be too small.

Compressor Discharge Valve Defective

1. Low head pressure.
2. Poor refrigeration.
3. Compressor excessively warm.
4. If compressor is stopped, pressure will equalize.

Expansion-Valve Needle Stuck Open

1. Poor refrigeration.
2. High head pressure if stuck partially open.
3. Low head pressure and high back pressure if stuck fully open.
4. Frosted or sweating suction line.
5. Hissing sound at expansion valve.
6. Impossible to adjust for higher or lower back pressures.
7. On methyl chloride and F-12 units, this may be caused by moisture freezing at seat of expansion valve.

8. Improper oil, freezing at seat.
9. Improper oil and high compressor temperatures may result in carbonization which may build up at the expansion valve, especially if filters are defective.

Expansion Valve Needle Stuck Closed

1. No refrigeration if shut tight.
2. Little refrigeration if stuck partially shut.
3. Evaporator will be pumped down so that low side will show an unduly low pressure.
4. A pinched liquid line, plugged filter, or closed hand valve will give the same symptoms.
5. On methyl chloride and F-12 units, this may be caused by moisture freezing in the expansion valves.

High Head Pressure

1. Air in system.
2. Excessive refrigerant charge.
3. Air or water going through condenser at too high a temperature.
4. If the unit is an air-cooled type, air circulation over condenser blocked.
5. If the unit is of the water-cooled type, the water may be turned off or restricted.
6. Rotation of motor reversed.
7. If a higher setting is used on the expansion valve, the head pressurewill be higher, and vice versa.

Unable to Adjust Valve

1. Refrigeration shortage.
2. Compressor valve broken or defective.
3. Partially plugged screens or filters.
4. Liquid line pinched almost closed.
5. Stoppage in fitting or restriction in liquid line.

6. Needle or seat eroded and leaky.
7. Oil-logged coil.
8. Ice in adjustment side of expansion valve.

Low Head Pressure

1. Refrigerant shortage.
2. Worn pistons.
3. Head or clearance gasket too thick.
4. Suction valve worn, split, or stuck.
5. Low setting on expansion valve.
6. Gasket between cylinders blown.

Suction Line and Drier Coil Frosted or Sweating

1. Expansion valve stuck open or leaky.
2. Needle seat eroded or corroded.
3. Valve set at too high a back pressure.
4. Ice or moisture in adjustment side of expansion valve.
5. Thermostat out of order or poorly located.
6. Fan not operating, so that air is not blown over coils.
7. No water, or water pump not operating to pass water over evaporator.

COMMON TROUBLES IN THERMOSTATIC EXPANSION-VALVE SYSTEMS WITH LOW-SIDE OR THERMOSTATIC CONTROL

Refrigerant Shortage

1. Continuous operation.
2. Low head pressure.
3. Poor refrigeration.
4. Warm or hot liquid line.
5. Evaporator coils not chilled throughout entire length.
6. Hissing at expansion line.

Poor Refrigeration

1. Heavy coating of ice or frost on evaporator coils.
2. Expansion valve set too high.
3. Refrigerant shortage.
4. Thermostat bulb placed where there is little change in coil temperature.
5. Thermostat bulb placed where it is in a cold pocket and not affected by average conditions.
6. Expansion valve set too low.
7. Pigtail of expansion valve improperly placed, so that maximum coil surface is not used.
8. Compressor valves defective, broken, or sticking.
9. Liquid line pinched.
10. Suction line pinched.
11. Strainer clogged.
12. Suction line too small for job.
13. Partially closed liquid- or suction-line hand valves.
14. Compressor too small for job.
15. Moisture in methyl chloride or F-12 refrigerant.

Compressor Discharge Valve Defective

1. Low head pressure.
2. Poor refrigeration.
3. When compressor is stopped, pressures equalize.

Expansion Valve Needle Stuck Open

1. Continuous operation.
2. Poor refrigeration.
3. High head pressure.
4. Hissing sound at expansion valve.
5. Moisture in methyl chloride or F-12 refrigerant.

Low Head Pressure

1. Shortage of refrigerant.
2. Worn pistons in compressor.

3. Warped, split, or stuck discharge valve.
4. Suction valve warped, split, or stuck.
5. Expansion-valve needle stuck wide open.
6. Gasket between cylinder blown.
7. Thermostatic bulb discharged.

Suction Line and Drier Coil Sweating or Frosted

1. Expansion-valve needle stuck open.
2. Expansion-valve needle or seat eroded and leaky.
3. Expansion valve set too high above cutout point.
4. Control-switch points fused together.
5. Low-side control switch locked in operating position.

Expansion Valve Needle Stuck Partially Closed

1. Little or no refrigeration.
2. The high-pressure safety cutout may trip.
3. Evaporator will be pumped down and show a low
   pressure at below cut-in temperature.
4. If the liquid line is plugged, the thermostatic bulb
   discharged, or the capillary tube pinched, the result will
   be the same as a needle stuck shut.

High Head Pressure

1. The high-pressure safety cutout may cause system to be
   stopped.
2. Air in the system.
3. Excessive refrigerant charge.
4. Air or water passing over condenser at too high a
   temperature.
5. If unit is water-cooled, flow may be restricted or
   turned off.
6. If a high setting is used on the expansion valve (resulting
   in high back pressure), the head pressure will be higher
   than if the suction pressure were low.

7. Rotation of fan for cooling condenser reversed.
8. Fan blades bent or air passing over condenser restricted.

Expansion Valve Cannot Be Adjusted

1. Oil-logged evaporator.
2. Shortage of refrigerant.
3. Compressor valve broken or stuck.
4. Partially plugged screen in filter.
5. Liquid line pinched.
6. Stoppage in fitting or restriction in liquid line.
7. Stop-valve seat dropped and sealing open.
8. Charge lost in valve bulb.
9. Valve bulb loosened in its cradle by frost action; not making proper contact.
10. Valve covered; not open to atmospheric conditions.
11. Valve in too cold a location.

CHAPTER 8

PROCESS PIPING USING PLASTICS

Process piping makes up a large part of pipefitters' and welders' work. Process piping can be generally defined as piping carrying substances, liquid or solid, used in manufacturing or processing of other materials or substances.

TYPES OF PIPING

PVC (Polyvinyl chloride) conforming to ASTM D-1784, Class 12454-B formerly designated as Type I, Grade 1.

PVC is the most frequently specified of all thermoplastic materials. It is used for chemical processing, industrial plating, chilled water distribution, deionized water lines, chemical drainage, and irrigation systems. PVC has high physical properties and resistance to corrosion and chemical attack by acids, alkalies, salt solutions, and many other chemicals. It is attacked, however, by polar solvents such as ketones, some chlorinated hydrocarbons, and aromatics. The maximum service temperature of PVC is 140°F. It has a design stress of 2,000 psig and the highest long term hydrostatic strength at 73°F of any of the major thermoplastics used for piping systems. PVC is joined by solvent cementing, threading, or flanging. The method used depends on the schedule pipe used, 40 or 80.

CPVC (Chlorinated polyvinyl chloride) conforming to ASTM D-1784, Class 23447-B, formerly designated Type IV, Grade 1.

CPVC has physical properties at 73°F similar to those of PVC, and its chemical resistance is similar to that of PVC. CPVC, a design stress of 2,000 psig and maximum service temperature of 210°F, has proven to be an excellent material for hot corrosive liquids, hot and cold water distribution, and similar applications above the temperature range of PVC. Like PVC, CPVC is joined by solvent cementing, threading, or flanging.

SOLVENT CEMENTING OF PVC AND CPVC PIPE AND FITTINGS

PVC and CPVC pipe and fittings are made with a high-gloss hard finish. A *primer* must be used when solvent cementing if a good joint is to be made. Primer penetrates and softens the hard surfaces of both the pipe and fittings. Follow the instructions furnished with the primer. After the primer has softened the pipe and fitting surfaces, apply the PVC and/or CPVC cement.

(PP) Polypropylene homopolymer, conforming to ASTM D 4101, Class PP110-B67154, formerly designated Type I.

PP is a member of the polyolefin family of plastics. PP has less physical strength than PVC, but it is chemically resistant to organic solvents as well as acids and alkalies. Polypropylene should not be used in contact with strong oxidizing acids, chlorinated hydrocarbons, and aromatics. PP is useful for salt-water disposal lines, crude oil piping, and low-pressure gas gathering systems. PP is an excellent material for laboratory and industrial drainage where mixtures of acids, bases, and solvents are involved. Polypropylene is joined by the thermo-seal fusion process, threading, or flanging.

PVDF (Kynar) Polyvinylidene Fluoride

PVDF is a high molecular weight fluorocarbon that has superior abrasion resistance, chemical resistance, dielectric properties, and mechanical strength. PVDF maintains these characteristics over a temperature range of −40°F. to 250°. PVDF is highly resistant to wet or dry chlorine, bromine and other halogens, most strong acids and bases, aliphatics, aromatics, alcohols, and chlorinated solvents. PVDF is not recommended for ketones or esters. PVDF is joined by the thermo-seal fusion process, threading, or flanging.

PIPE PREPARATION

1. Cutting

Plastic pipe can be cut easily using a power or handsaw, circular or band saw. When a handsaw is used, it must be used with a miter box to insure square cuts. Tubing cutters are designed for quick, clean cuts through plastic pipe. When tubing cutters are used, the cutters should have cutting wheels especially designed for cutting plastic pipe. The cutter shown in Fig. 8-1 has a quick-acting slide mechanism and is made in five sizes to cut plastic pipe from $1/8$ in. through 6 inches O.D. (outside diameter). The cutter shown in Fig. 8-2 also has a compound ratchet lever mechanism and will cut pipe from $1/8$ in. through $2^3/8$ in. O.D.

2. Deburring and Beveling

All burrs, chips, filings, etc. should be removed from both the pipe I.D. (inside diameter) and O.D. Use a knife, half-round file or a deburring tool (Fig. 8-3) to remove all burrs. Pipe should be beveled to approximately the angle shown in Fig. 8-4.

POLYPROPYLENE PIPING AND FITTINGS

Pipe Joining

The thermal bonding joining method which is detailed herein applies to all NIBCO/Chemtrol polypropylene and Kynar pressure piping systems, including molded socket fittings and socket type valve connections. This process involves the application of regulated heat uniformly and simultaneously to pipe and fitting mating surfaces so that controlled melting occurs at these surfaces. All recommendations and instructions presented herein for thermal bonding are based upon the use of a Thermal-Seal fusion tool for applying uniform heat to pipe and fittings.

Courtesy Ridge Tool Co.

Fig. 8-1. A quick-acting plastic-pipe cutter.

Courtesy Ridge Tool Co.

Fig. 8-2. A compound-lever plastic-pipe cutter.

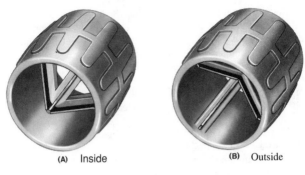

(A) Inside   **(B)** Outside

Fig. 8-3. A deburring tool. Courtesy Ridge Tool Co.

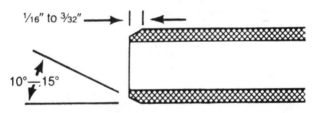

Fig. 8-4. Pipe beveled to correct angle.

JOINING EQUIPMENT AND MATERIALS

Vise
Cutting tools
Deburring tool
Thermal-Seal tool
Electric Model DD-1 with $1/2$" to 2" tool pieces
or
Electric Model DD-2 with $1/2$" to 6" tool pieces.
A Thermo-Seal joining kit for pipe sizes $1/2$" to 4" as shown
in Fig. 8-5.

Electric Model tools are available for making socket fusion joints. They are the preferred socket fusion tools because the thermostatically controlled heat source automatically maintains fusion temperatures within the recommended range.

Electric Model DD-1 is electrically heated and thermostatically controlled. This tool is designed to join polypropylene and PVDF pipe, valves, and fittings in sizes 1/2" through 2". The unit operates on 110 v ac (6.7 amps; 800 watts) electricity and is fitted with ground wires.

The Electric Model DD-2 is also electrically heated and thermostatically controlled. It is used to join polypropylene pipe and fittings in sizes 1/2" through 6". This unit operates on 110 v ac, 15 amps; 1650 watts.

*CAUTION:*Thermal bonding and fillet welding involve temperatures in excess of 540° F. Severe burns can result from contacting equipment or molten plastic material at or near these temperatures.

PIPE PREPARATION

Cutting

Polypropylene and Kynar can be easily cut using a power or handsaw, circular, or band saw. For best results use a fine-toothed blade, 16 or 18 teeth per inch. A circumferential speed of 6,000 ft. per minute is recommended for circular saws; a speed of 3,000 ft. per minute is suitable for band saws. It is important that cuts must be square. To insure square cuts when sawing, a miter box must be used. Pipe or tubing cutters can be used, but the cutting wheel must be specifically designed for plastics.

Deburring and Beveling

All burrs, chips, and filings should be removed from both the pipe I.D. (inside diameter) and O.D. (outside diameter) before joining. A knife, half round file, or a deburring tool can be used. The deburring tool shown in Fig. 8-3 will remove burrs and also bevel the pipe. Pipe ends should be beveled to approximately the angle shown in Fig. 8-4 for ease of socketing and to minimize the chances of wiping melt material from the I.D. of the fitting as the pipe is socketed.

FITTING PREPARATION

Use a clean dry cotton rag to wipe away all loose dirt and moisture from the I.D. and O.D. of the pipe end and the I.D. of the fitting. ***DO NOT ATTEMPT TO THERMAL BOND WET SURFACES.***

In order to provide excess material for fusion bonding, polypropylene and Kynar components are manufactured to socket

Courtesy CHEMTROL, A Division of NIBCO, Inc.
Fig. 8-5. A Thermo-Seal joining kit for pipe sizes 1/2" to 4".

dimensions in which the socket dimensions are smaller than the pipe O.D. It should not be possible to easily slip the pipe into the fitting socket past the initial socket entrance depth, and in no case should it ever be possible to bottom the pipe in the socket prior to fusion. If a fitting socket appears to be oversize, it should not be used.

Setting Up the Thermo-Seal Tool

1. Install the male and female tool pieces on either side of the Thermo-Seal tool and secure with set screws (Fig. 8-6).

2. Insert the electrical plug into a grounded 110 v electrical source and allow the tool to come to the proper operating temperature (Fig. 8-7). The tool temperature is read directly from

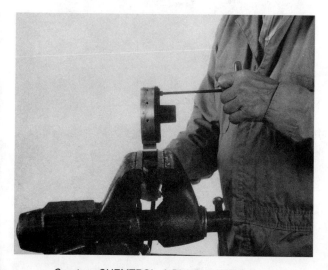

Courtesy CHEMTROL, A Division of NIBCO, Inc.
Fig. 8-6. Install the male and female tool pieces.

Fig. 8-7. Insert the grounded plug into a grounded 110 v electrical source.

the mounted temperature gauge, and tool temperature can be adjusted by turning the thermostat adjustment screw with a screwdriver. (Counterclockwise to raise the temperature and clockwise to lower the temperature.) One turn of the screw will give approximately a 25° temperature change.

Good thermal bonded joints can be made only when the Thermo-Seal tool is operating at the proper temperature and only when the length of time that the pipe and fittings remain on the heated tool pieces does not exceed those times recommended for the particular size of pipe and fitting to be joined, as shown in Table 8-1. The heating times shown in Table 8-1 should begin after pipe and fitting have been completely placed in position on the heat tool.

Excessive temperatures and excessive heating times will result in excessive melting at and below the surfaces of the fitting socket I.D. and pipe O.D. When the pipe is inserted into the fitting socket, excessive melt material needed for thermal bonding will be scraped from the socket wall and into the fitting waterway, and the resulting joint will be defective.

Making Thermal Bonded Joints

1. Place the proper size depth gauge over the end of the pipe as shown in Fig. 8-8.

2. Attach the depth gauging clamp to the pipe by butting the clamp up to the end of the depth gauge and locking it into place. Then remove the depth gauge, Fig. 8-9.

3. Simultaneously place pipe and fitting squarely and fully on the heat tool pieces so that the I.D. of the fitting and the O.D. of the pipe are in contact with the heating surfaces. Care should be taken to insure that the pipe and fitting are not cocked when they are inserted on the tool pieces, as shown in Fig. 8-10.

4. Hold the pipe and fitting on the tool pieces for the prescribed amount of time. During this time a bead of melted material will appear around the complete circumference of the pipe at the entrance of the tool piece, Fig. 8-11.

5. Simultaneously remove the pipe and fitting from the tool pieces and immediately insert the pipe, squarely and fully, into the socket of the fitting. Hold the completed joint in place and avoid relative movement between components for at least 15 seconds, Fig. 8-12. Once a joint has been completed, the clamp can be removed and preparation for the next joint can be started, Fig. 8-13.

6. The surfaces of the female and male tool surfaces are teflon coated to prevent sticking of the hot plastic, Fig. 8-14.

Table 8-1. Recommended HEATING TIME — Polypropylene

Pipe Size	Heating Time at 540°F
1/2"	10–15 sec.
3/4"	13–18 sec.
1"	14–20 sec.
1 1/2"	15–20 sec.
2"	20–25 sec.
3"	25–30 sec.
4"	25–35 sec.
6"	35–50 sec.

Recommended HEATING TIME — PVDF

Pipe Size	Heating Time at 565°F
1/2"	20–25 sec.
3/4"	20–25 sec.
1"	25–30 sec.
1 1/2"	30–45 sec.
2"	30–45 sec.
3"	30–45 sec.
4"	45–60 sec.
6"	60–75 sec.

Courtesy CHEMTROL, A Division of NIBCO, Inc.

Table 8-2. Recommended Rod Sizes

Pipe Sizes	Rod Sizes	Number of Passes
1/2"–3/4"	3/32"	3
1"–2"	3/32"	3
2 1/2"–4"	1/8"	3
6"–8"	1/8" or 5/32"	5
10"–12"	5/32" or 3/16"	5

Courtesy CHEMTROL, A Division of NIBCO, Inc.

Courtesy CHEMTROL, A Division of NIBCO, Inc.

Fig. 8-8. Place the proper size depth gauge over the end of the pipe.

Courtesy CHEMTROL, A Division of NIBCO, Inc.

Fig. 8-9. Attach the depth gauging clamp to the pipe.

Courtesy CHEMTROL, A Division of NIBCO, Inc.

Fig. 8-10. Simultaneously place pipe and fitting squarely and fully on the heat tool pieces.

Courtesy CHEMTROL, A Division of NIBCO, Inc.

Fig. 8-11. Hold the pipe and fitting on the tool pieces for the prescribed amount of time.

Courtesy CHEMTROL, A Division of NIBCO, Inc.

Fig. 8-12. Simultaneously remove the pipe and fitting from the tool pieces.

Courtesy CHEMTROL, A Division of NIBCO, Inc.

Fig. 8-13. Once a joint has been completed, the clamp can be removed.

Courtesy CHEMTROL, A Division of NIBCO, Inc.

Fig. 8-14. The surfaces of the female and male tool pieces are teflon coated to prevent sticking of the hot plastic.

It is important that the tool pieces be kept as clean as possible. Any residue left on the tool pieces should be removed immediately by wiping with a cotton cloth.

CAUTION: HOT PLASTIC MATERIAL CAN CAUSE SEVERE BURNS. AVOID CONTACT WITH IT.

The five basic procedures for making good thermal bonded joints are:

1. The tool must be operated at the proper temperature.
2. The fitting must be slipped *squarely* onto the male tool while the pipe is simultaneously inserted into the female tool.
3. The fitting and pipe must *not* remain on the heat tool for an excessive period of time. Recommended heating times must be followed.

4. The pipe must be inserted *squarely* into the fitting socket *immediately* after removal from the heated tools.

5. The Thermo-Seal tool must be kept clean at all times.

Pressure Testing

The strength of a thermal bonded joint develops as the material in the bonded area cools. One hour after the final joint is made, a thermal bonded piping system can be pressure tested up to 100% of its hydrostatic pressure rating.

CAUTION: AIR OR COMPRESSED GAS IS NOT RECOMMENDED AND SHOULD NOT BE USED AS A MEDIUM FOR PRESSURE TESTING OF PLASTIC PIPING SYSTEMS.

REPAIRING THERMOPLASTIC PIPE JOINTS

The most common method for repairing faulty and leaking joints is hot gas welding at the fillet formed by the junction of the fitting socket entrance and the pipe. Hot gas welding (which is similar to gas welding with metals except that *hot gas* is used for melting *instead of a direct flame*) consists of simultaneously melting the surface of a plastic filler rod and the surfaces of the base material in the fillet area while forcing the softened rod into the softened fillet.

Welding with plastics involves only surface melting because plastics, unlike metals, must never be "puddled." The resulting weld is not as strong as the parent pipe and fitting material, therefore, fillet welding is recommended for minor leaks only.

Pipe and fittings must always be the same type; dissimilar pipe and fittings must never be joined. *Polypropylene pipe and fittings must never be joined to Kynar pipe and fittings.*

Welding Tools And Materials

Hot gas welding torch with 400–550 watt heating element and pressure regulator equipped with pressure gauge. A hot gas welding torch is shown in Fig. 8-15.

Filler rod

Emery cloth

Cotton rags

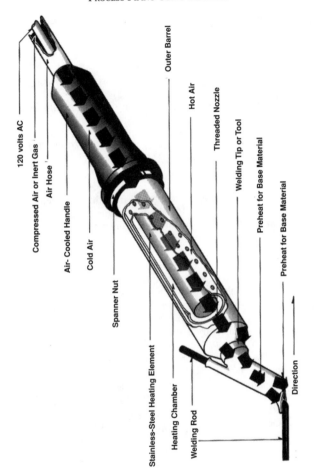

Courtesy KAMWELD Products Co. Inc.

Fig. 8-15. A hot gas welding torch.

Cutting pliers
Hand grinder (optional)
Nitrogen or other inert gas (Polypropylene or Kynar only)
Source of compressed air

Weld Area Preparation

Wipe all dirt, oil and moisture from the joint area. A mild solvent may be needed to remove oil.

CAUTION: MAKE SURE THAT ALL LIQUID HAS BEEN REMOVED FROM THE PORTION OF THE PIPING SYSTEM WHERE THE WELD IS TO BE MADE.

Welding Faulty Joints

1. Remove residual solvent cement from the weld area using emery cloth (Fig. 8-16). When welding threaded joints, a file can be used to remove threads in the weld area.

2. Wipe the weld area free of dust, dirt, and moisture (Fig. 8-17).

3. Determine the amount of the correct filler rod (Table 8-2) necessary to make one complete pass around the joint by wrapping the rod around the pipe to be welded. Increase this length enough to allow for handling the rod at the end of the pass (Fig. 8-18).

4. Make about a 60° angular cut on the lead end of the filler rod (Fig. 8-19). This will make it easier to initiate melting and will insure fusion of the rod and base material at the beginning of the weld.

5. Welding temperatures vary for different thermoplastic materials (500°F–550°F for PVC and CPVC, 550°F–600°F for PP, 575°F–600°F for Kynar). Welding temperatures can be adjusted for the various thermoplastic materials as well as any for desired welding rate by adjusting the pressure regulator (which controls the gas flow) between 3 and 8 psig (Fig. 8-20).

6. With air or an inert gas flowing through the welding torch (Fig. 8-15), insert the electrical plug for the heating element into an appropriate electrical socket to facilitate heating of the gas, and wait approximately 7 minutes for the welding gas to reach the proper temperature (Fig. 8-21).
   CAUTION: THE METAL BARREL OF THE HEATING TORCH HOUSES THE HEATING ELEMENT. IT CAN ATTAIN EXTREMELY HIGH TEMPERATURES. AVOID CONTACT WITH THE BARREL AND DO NOT ALLOW IT TO CONTACT ANY COMBUSTIBLE MATERIALS.

Courtesy CHEMTROL, A Division of NIBCO, Inc.

Fig. 8-16. Remove residual solvent cement from the weld area using emery cloth.

Courtesy CHEMTROL, A Division of NIBCO, Inc.

Fig. 8-17. Wipe the weld area clean of dust, dirt, and moisture.

Courtesy CHEMTROL, A Division of NIBCO, Inc.

Fig. 8-18. Determine the amount of correct filler rod needed.

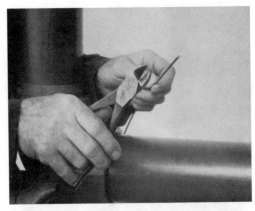

Courtesy CHEMTROL, A Division of NIBCO, Inc.

Fig. 8-19. Make a 60° cut on the filler rod.

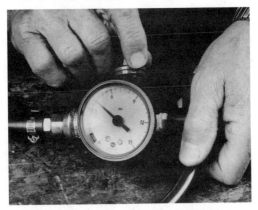

Courtesy CHEMTROL, A Division of NIBCO, Inc.

Fig. 8-20. Adjust the regulator for proper welding temperature.

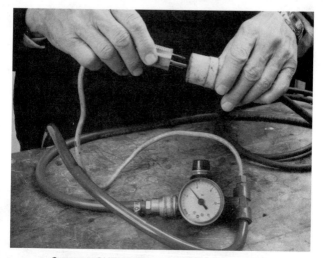

Courtesy CHEMTROL, A Division of NIBCO, Inc.

Fig. 8-21. Insert the electrical plug for the heating element into a 110 v AC receptacle.

7. Place the leading end of the filler rod into the fillet
    formed by the junction of the pipe and fitting socket
    entrance (Fig. 8-22). Hold the filler rod at an angle of 90°
    to the joint for PVC, CPVC, and Kynar; 75° to the joint
    for polypropylene. Preheat the surfaces of the rod and
    base materials at the weld starting point by holding the
    welding torch steady at approximately $1/4$ to $3/4$ inches
    from the weld starting point and directing the hot gas in
    this area until the surfaces become tacky. While
    preheating, move the rod up and down slightly so that the
    rod slightly touches the base material. When the surfaces
    become tacky, the rod will stick to the base material.

8. Advance the filler rod forward by applying a slight pressure to the rod. Simultaneously apply even heat to the surfaces of both the filler rod and base material by moving the torch with a fanning or arcing motion at a rate of about 2 cycles per second. The hot gas should be played equally on the rod and base material along the weld line for a distance of about 1/4 inch from the weld point as seen in Fig. 8-23.

IMPORTANT: If charring of the base or rod material occurs, move the tip of the torch back slightly, increase the fanning frequency, or increase the gas flow rate. If the rod or base materials do not melt sufficiently, reverse the above corrective measures. Do not apply too much pressure to the rod because this will tend to stretch the weld bead causing it to crack and separate after cooling.

Courtesy CHEMTROL, A Division of NIBCO, Inc.

Fig. 8-22. Place the leading end of the filler rod into the fillet formed by the junction of the pipe and fitting socket.

Courtesy CHEMTROL, A Division of NIBCO, Inc.

Fig. 8-23. Advancing the filler rod forward.

9. When welding large-diameter pipe, three passes may be required. The first bead should be deposited at the bottom of the fillet, and subsequent beads should be deposited on each side of the first bead as seen in Fig. 8-24.

10. Since the starting point for a plastic weld is frequently the weakest part of the weld, always terminate a weld by lapping the bead on top of itself for a distance of 3/8 to 1/2 inches as shown in Fig. 8-25. Never terminate a weld by overlapping the bead side by side.

11. Properly applied plastic welds can be recognized by the presence of small flow lines or waves on both sides of the deposited bead. This indicates that sufficient heat was applied to the surfaces of the rod and base materials to effect adequate melting and that sufficient pressure was applied to the rod to force the rod melt to fuse with base material melt. If insufficient heat is used when welding PVC, CPVC, or Kynar, the filler rod will appear in its

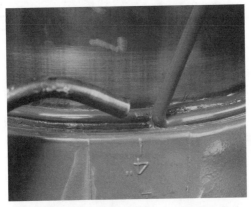

Courtesy CHEMTROL, A Division of NIBCO, Inc.

Fig. 8-24. Three passes may be needed on large pipe.

Courtesy CHEMTROL, A Division of NIBCO, Inc.

Fig. 8-25. Terminating a plastic weld.

original form and can easily be pulled away from the base material. Excessive heat will result in a brown or black discoloration of the weld. In the case of polypropylene, excessive heat will result in a flat bead with oversized flow lines.

12. Always unplug the electrical connection to the heating element and allow the welding torch (gun) to cool before shutting off the gas to the gun.

Portions of the preceding text have been reproduced from the Thermoplastic Piping technical Manual by permission of CHEMTROL, a Division of NIBCO, Inc., in order to instruct the reader in the proper method of thermo-bonding polypropylene and Kynar pipe and fittings and in thermal welding of certain types of plastic piping.

CHAPTER 9

GROOVED-END—PLAIN-END PIPING SYSTEMS

Every pipe fitter and sprinkler fitter should become familiar with the use of grooved-end and plain-end piping. There are two types of grooving: cut grooving and roll grooving. Grooved piping has been widely used in fire protection systems for many years, and technological advances (and economic conditions) now make the use of grooved-end piping systems desirable in many applications. Grooved-end piping can be used in both metallic and plastic piping systems. Installation techniques are similar, so because of space limitations, only grooved-end and plain-end *metallic* piping is explained here.

The Victaulic® Company of America is the originator and developer of the grooved piping system of joining pipe and fittings mechanically. The concept is based on a rugged ductile iron housing which grips into the pipe, a synthetic pressure-responsive rubber gasket to seal the system, and nuts and bolts to secure the components into a unified joint. Each joint is a union. The removal of two couplings permits removal of a section of pipe, a fitting, or a valve for cleaning, repair, or replacement.

CUT-GROOVING METALLIC PIPE

Utilization of the cut-grooved method is based on the proper preparation of a groove in the pipe end to receive the housing key. The groove serves as a recess in the pipe with ample depth for secure engagement of the housing yet with ample wall thickness for full pressure rating. Fig. 9-1 is a cutaway view of a Victaulic®

FLEXIBLE

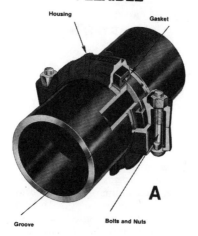

RIGID

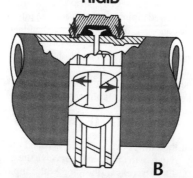

Courtesy Victaulic® Company of America.

Fig. 9-1. Cutaway view of a Victaulic® grooved piping coupling.

cut-groove coupling showing how the gasket and the bolts are installed. Standard bolts for most couplings are heat treated oval neck track head bolts. The oval neck fits into oval-shaped holes in the housing, permitting the nuts to be tightened from one side with a ratchet wrench.

There are two types of couplings: one for rigid pipe installations, the other for flexible piping installations. Rigid couplings create a rigid joint, useful for risers, mechanical rooms, and other areas where flexibility is not desired. Flexible couplings provide allowance for controlled pipe movement—expansion, contraction and/or deflection—to absorb movement from thermal changes, settling, or seismic action and to dampen noise and vibration.

Pipe can be delivered to the job with the grooves already cut, or the grooves can be cut on the job. Tools designed for on-the-job use will groove pipe ranging from $3/4$" I.D. (inside diameter) to 24" I.D. Most are designed for power drive operation; smaller size groovers can be driven manually or by a power drive. The grooving tool shown in Fig. 9-2 will cut-groove metallic pipe ranging from 2 in. I.D. (inside diameter) through 8 in. I.D. The tool shown in Fig. 9-3 will cut-groove metallic pipe in sizes from 8" I.D. through 24" I.D.

A roll-grooving tool mounted on a power drive is shown in Fig. 9-4. Pipe can be cut to length, ready for grooving, by use of standard pipe cutters or oxy-gas torches.

Grooved-end fittings are made for use with grooved-end piping. Tees, elbows, ($90°$; $45°$; $22^1/2°$; $11^1/4°$), street ells, reducer ells, tees, reducer tees, crosses, wyes, flanged adapters, threaded adapters, and eccentric and concentric reducers are some of the many standard pattern fittings shown in Fig. 9-5 that are made for use with grooved piping systems. Fittings for American Water Works Association (AWWA) sized pipe are available. A wide assortment of grooved-end valves are made for use with grooved piping systems. Various types of butterfly valves are shown in Fig. 9-6.

Steps in assembly of couplings for connecting a $90°$ elbow in a grooved piping system are shown in Fig. 9-7. Grooved piping systems are used extensively for process piping in industrial plants and mining operations, in oilfields and offshore oil producing

Fig. 9-2. Cut-grooving tool for 2" I.D. through 8" I.D. pipe.

Fig. 9-3. Cut-grooving tool for 8" I.D. through 24" I.D. pipe.

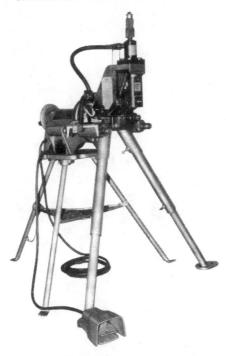

Fig. 9-4. A roll-grooving tool mounted on a power drive.

facilities, in power plants, and in municipal sewage and water treatment plants.

Flanges can also be installed in grooved piping systems as shown in Fig. 9-8.

Grooved piping systems are also used in fire protection systems for underground and above ground mains, wet and dry standpipes, branch lines for sprinkler heads, and piping conducting Halon or other fire extinguishing materials.

Courtesy Victaulic® Company of America.

Fig. 9-5. Victaulic® fittings are made for every piping need.

One of the principal advantages in the use of grooved piping systems is that no oil or pipe dope is used in assembling the components of the system, and contamination of the piping from these products is eliminated.

Plain-End Piping Systems For Steel Pipe

The Victaulic® plain-end method can be used to simplify installation, maintenance and repairs of piping systems. No special pipe end preparation is required. Pipe cuts should be square.

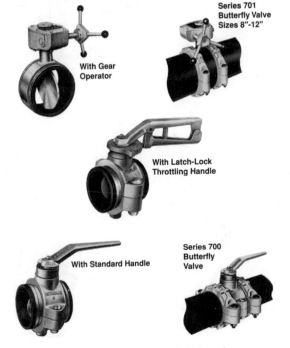

With Gear Operator

Series 701 Butterfly Valve Sizes 8"-12"

With Latch-Lock Throttling Handle

With Standard Handle

Series 700 Butterfly Valve

Fig. 9-6. Various types of Victaulic® butterfly valves.

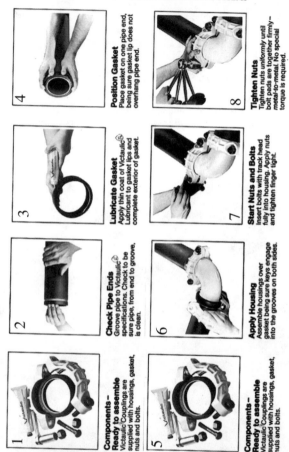

4 — Position Gasket
Place gasket on one pipe end, being sure gasket lip does not overhang pipe end.

8 — Tighten Nuts
Tighten nuts uniformly until bolt pads are together firmly—metal-to-metal. No special torque is required.

3 — Lubricate Gasket
Apply thin coat of Victaulic® Lubricant to gasket lips and complete exterior of gasket.

7 — Start Nuts and Bolts
Insert bolts with track head fully into housing. Apply nuts and tighten finger tight.

2 — Check Pipe Ends
Groove pipe to Victaulic® specifications. Check to be sure pipe, from end to groove, is clean.

6 — Apply Housing
Assemble housings over gasket, being sure keys engage into the grooves on both sides.

1 — Components—Ready to assemble
Victaulic Couplings are supplied with housings, gasket, nuts and bolts.

5 — Components—Ready to assemble
Victaulic Couplings are supplied with housings, gasket, nuts and bolts.

Courtesy Victaulic® Company of America.

Fig. 9-7. Steps in assembly of a 90° elbow into a piping run.

Courtesy Victaulic® Company of America.

Fig. 9-8. Installing a flange on cut-grooved pipe.

Victaulic® Plain-End Couplings can also be used with beveled-end pipe. The plain-end system uses grips in the couplings which bite into the pipe ends and secure them. These couplings are primarily designed for use on *standard weight (schedule 40) steel pipe.* They are not intended for use on plastic pipe, plastic coated pipe, or brittle pipe such as asbestos cement or cast iron. They are not intended for use on pipe with a surface hardness greater than 150 Brinell.

Courtesy Victaulic® Company of America.

Fig. 9-9. A Victaulic® plain-end coupling.

Plain-end fittings are available in standard patterns common to the trade. Grooved-end fittings must not be used with Victaulic® Plain-End Couplings shown in Fig. 9-9.

The Victaulic® Plain-End Piping System used with the correct Victaulic® fittings and valves is listed for fire protection services by Underwriters Laboratories.

We have explained only Victaulic® fittings for *metallic* piping in this chapter. Victaulic® fittings are also available for use with plain-end *high density polyethylene or polybutylene* pipe.

CHAPTER 10

LEARNING TO USE AN INSTRUMENT LEVEL

A skilled pipefitter should have a working knowledge of elevations. Especially on new construction, sleeves and boxes must often be set at exact points in reinforced poured-concrete walls and footings. Underground piping may have to be installed at an exact level to meet job specifications.

An instrument level is a telescope (Fig. 10-1) mounted on a base equipped with leveling screws (Fig. 10-2). The base is screwed onto a tripod (Fig. 10-3) when the instrument is in use. The instrument level can be rotated a full 360° horizontally, and when properly adjusted, the level-indicating bubble will show that the instrument is level at any point of the compass.

Adjusting an instrument level must be done very carefully in order to avoid damage to the instrument. After mounting the instrument on the tripod, swing the instrument so that the barrel is directly over two opposing screws (A and C, Fig. 10-2). Both screws should be turned equally and simultaneously in the same direction until the leveling bubble is centered. Then swing the barrel 180 degrees. The bubble should still be centered. Then swing the barrel 90° until the barrel is directly over the other two opposing screws (B and D, Fig. 10-2). Adjust these screws until the bubble is centered. Then reverse the instrument, turn 180° again, still over B and D; the bubble should remain centered. Swing again to A and C and re-adjust these until the bubble is again centered. It may be necessary to repeat this process several times before the instrument is in a level position when pointed in any direction. *Do not over-tighten the adjusting screws.* If the adjusting screws are

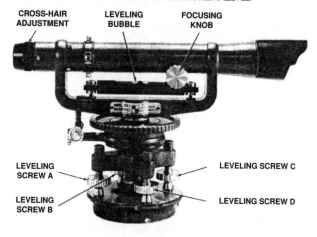

Courtesy David White Instruments

Fig. 10-1. An instrument level.

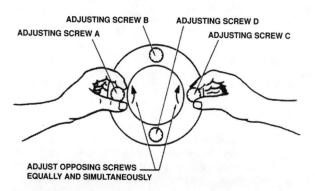

Fig. 10-2. Correct way to adjust leveling screws.

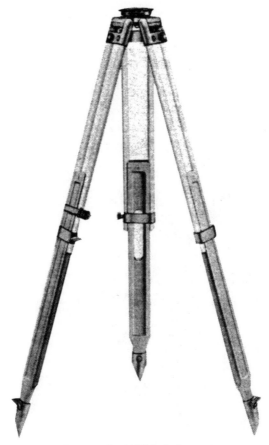

Courtesy David White Instruments

Fig. 10-3. A tripod used with an instrument level.

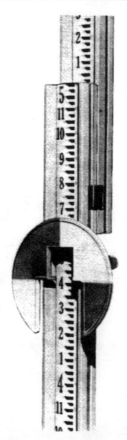

Courtesy David White Instruments

Fig. 10-4. An engineer's rod.

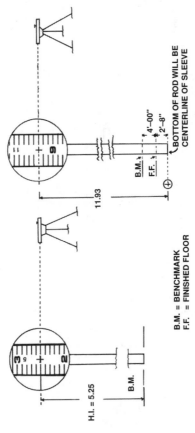

Fig. 10-5. Using an instrument level to mark correct elevation.

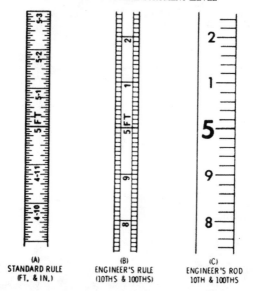

Fig. 10-6. A standard rule, an engineer's rule, and an engineer's rod.

over-tightened, the bed-plate of the instrument can be damaged thereby preventing leveling of the instrument altogether.

CAUTION: If, after repeated tries, the bubble will not remain centered when the instrument is reversed, the instrument is faulty and should be repaired.

One of the first steps when a new building is started is to establish a "bench" mark. This is usually done by a mechanical engineer. The bench mark is a fixed point from which all building measurements are taken, the top of an iron manhole in a street, for example. As the building goes up, the benchmark can be transferred to other locations where it will be visible from any point. This point is then given a number, and all vertical measurements of the building are based on this number. Some architects and mechanical engineers use a number based on the height of the area above sea level: 745.00 for instance. Others may use the number 100.00. The *metric* system of 10ths and 100ths is used because it is more accurate and easier to use than feet and inches. The number used

HELPFUL CONVERSIONS AND EQUIVALENTS

Approximate Conversions from Metric Measures

Symbol	When You Know	Multiply by	To Find	Symbol
		LENGTH		
mm	millimeters	0.04	inches	in
cm	centimeters	0.4	inches	in
m	meters	3.3	feet	ft
m	meters	1.1	yards	yd
km	kilometers	0.6	miles	mi

Approximate Conversions from Metric Measures

Symbol	When You Know	Multiply by	To Find	Symbol
		LENGTH		
in	inches	2.5	centimeters	cm
ft	feet	30	centimeters	cm
yd	yards	0.9	meters	m
mi	miles	1.6	kilometers	km

Fractional Inches	1/64	1/32	1/16	1/8	1/4	1/2	3/4
Decimal Inches	.016	.031	.063	.125	.25	.50	.75

Fig. 10-7. English-Metric conversions.

is not important since it is merely a reference point from which all vertical measurements are made.

As an example, if the bench mark is established at 747.00, plans may show a first floor, (ground level) finished height of 743.00, a second floor level of 753, and a basement floor level of 731.00. Numbers *greater* than the bench mark show elevations above the bench mark. The finished second floor will be 6 feet above the bench mark. (747.00 + 6 = 753.00). Numbers *lower* than the bench mark show elevations below the bench mark. The basement floor level will be 16 feet below the bench mark. (747.00 − 731.00 = 16).

When an instrument level is used, it must be set up at a point from which both the bench mark and the elevation to be located are visible. An engineer's rod (Fig. 10-4) marked in 10ths and 100ths of a foot (metric scale) is set up on the bench mark, and with the instrument leveled and pointed at the rod, a reading is taken. In the example shown in Fig. 10-5, the reading is 5.25. This is the *H.I. or height of the instrument above the bench mark.*

Using the above numbers, a pipefitter must set a sleeve for process piping in a form for a foundation wall. The sleeve must be centered at 2'8" below the finished first floor. How can he pinpoint this location?

First, he will set up his instrument level and take a reading on an engineer's rod placed on the bench mark. This reading shows 5.25, giving an H.I. of 5.25. It is very important that the rod be held straight when it is held to obtain a reading. If the rod is leaning toward the instrument, the reading will be high; if leaning away, the reading will be low.

Every fitter working with elevations will need an engineer's rule. A 6 ft. folding engineer's rule is a conversion rule with feet and inches on one side and metric scale on the other side. This is shown in Fig. 10-6. With the thumb nail placed on the feet and inches side of the engineer's rule, turn the rule over to see the equivalent metric scale. Example: 2'8" equals 2.68 metric feet. Subtracting 2.68 from 743.00 (finished first floor) shows that the elevation of the center line of the sleeve will be 740.32. We know that the H.I. of the instrument is 5.25 above the bench mark. We also know that the centerline of the sleeve must be 6'8" below the bench mark. If the engineer's rod is taken to the form location and lowered until the cross-hairs of the instrument read 11.93 (5.25 + 6.68) and the form is marked at the bottom of the rod, this mark will be the centerline of the sleeve.

The elevation of each point is shown in Fig. 10-5.

The conversion guides shown in Fig. 10-7 are helpful in converting measurements to and from metric.

CHAPTER 11

PNEUMATIC CONTROL SYSTEMS

The installation and servicing of pneumatic control systems falls within the scope of work of pipefitters. If I were starting out to learn a trade all over again, I would specialize in the installation and servicing of pneumatic controls as soon as I had completed my apprenticeship or equivalent training as a pipefitter. The installation of pneumatic controls is a very specialized field—and a very rewarding one. I'll explain how and where training in this field is available at the end of this chapter.

Although there are many uses for pneumatic controls in manufacturing facilities, process piping, the aerospace industry, fire systems control, theme parks, and numerous others, limited space will not permit the mention of all the uses of these automatic controls. In this chapter we will be more concerned with pneumatic systems to control heating/cooling and ventilating systems. The work involved includes the installation and servicing of air compressors, piping from compressors to pneumatically operated thermostats, valves, dampers, humidistats, and auxiliary devices. An air compressor specially designed for use with pneumatic controls is shown in Fig. 11-1.

FUNDAMENTALS OF PNEUMATIC CONTROL

A basic control system consists of a power supply, supply line, controller, output line, and the controlled device. A simplified basic control system is shown in Fig. 11-2. The air compressor (power supply) supplies pressure to the room thermostat (controller) which sends an output pressure signal to the damper actuator to position the damper (controlled device).

Fig. 11-1. An air compressor especially designed for use with pneumatic controls.

POWER SOURCE

Compressed air at a constant pressure of 15 to 25 psig (pounds per sq. in. gauge) is used as the power source. The power supply is the most important component of the control system. If it fails, the entire system will fail. Pneumatic power air must be clean, dry, and free of oil, and must be at a constant pressure. A refrigerated air dryer (Fig. 11-3) will provide dry air for the control system.

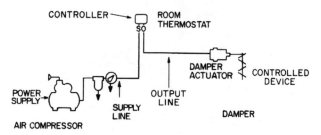

Fig. 11-2. A basic pneumatic control system.

SAFE OPERATION

Safe conditions are essential before mechanical equipment may be allowed to operate. "Fail-Safe" pneumatic controls can prevent damage caused by freezing or overheating. Spring-loaded actuators and valves cause these devices to return to a fail-safe position in the event of temperature control failure.

SYSTEMS OPERATION

All automatic control systems function on the "cause and effect" principle. This means that *every component in the system has an effect on the other components.*

In Fig. 11-4 the thermostat (Tr) measures the temperature of the air surrounding it. As the temperature rises, the thermostat causes a reduction of the heat (Hi) being supplied to the room which allows the room temperature (Rt) to stop rising or to drop. This affects the thermostat so that it adjusts its influence on heat input until a balance between heat input and heat loss is established, resulting in stabilized room temperature. Thus, one change is dependent on another change, and a "closed loop" system has been established.

A Closed Loop System is the arrangement of components to allow system feedback.

Example: a heating unit, valve, and thermostat arranged so that each component affects the other and can react to it. The relationship of "cause and effect," the interdependence of one thing upon another, is called "feedback." Feedback makes true automatic control possible.

Fig. 11-3. A refrigerated air dryer for use with pneumatic control systems.

There are three basic elements which must be considered when putting together a closed loop system. They are:

1. *The Control Agent.* The source of energy supplied to the system can be either *hot or cold,* such as steam, hot water, heated air, chilled water, electrical current, or refrigerant.
2. *The Controlled Device.* The instrument that receives the output signal from the controller and regulates the flow of

the Control Agent. It is functionally divided into two parts:

a. Actuator: receives the output signal and converts it into force.

b. Regulator: valve body or damper which regulates the flow of the Control Agent.

Note: A valve or damper can be either *normally open (NO)* or *normally closed (NC)* to regulate the flow of the control agent. The type to be used is chosen primarily for "failsafe" operation.

3. *Controller Action:* the controller that is furnished may be either direct acting or reverse acting. The controller will allow the system to balance itself.

   a. *Direct Acting:* the output signal changes in the same direction in which the controlled or measured variable changes. An increase in the controlled variable results in an increased output signal.

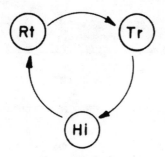

Tr = ROOM THERMOSTAT
Hi = HEAT INPUT
Rt = ROOM TEMPERATURE

Fig. 11-4. Schematic of components of a closed loop system.

b. *Reverse Acting:* The output signal changes in the opposite direction from that in which the controlled or measured variable changes. An increase in the controlled variable results in a decreased output signal.

The closed loop system is the most commonly used system. A typical closed loop system with a *two-position* (either ON or OFF) pneumatic controller is shown in Fig. 11-5. The closed loop system (Fig. 11-6) uses a *proportional* pneumatic controller, which holds the valve open to admit just enough of the heating medium to the heating coil, maintaining room temperature at or near the desired level.

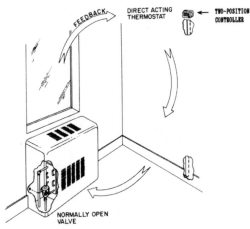

Typical Closed Loop System with two-position pneumatic controller. Thermostat (controller) measures room temperature and opens valve (controlled device) when temperature falls to lower limit of differential. This admits steam to terminal unit which in turn heats the air in the room. Then, room temperature rises providing feedbck to the thermostat which closes valve when temperature reaches top limit of differential.

Fig. 11-5. A closed loop system with a two-position controller.

ACTUAL OPERATING SEQUENCE

To put everything we've learned in perspective, let's go to Fig. 11-5 to see exactly what happens when the room temperature drops.

When the temperature drops below the set point of the thermostat, the thermostat (controller) opens the steam valve to the heating unit, which, in turn, heats the air in the room. Then, room temperature rises, providing feedback to the thermostat. The thermostat closes the steam valve when room temperature reaches the top limit of the differential.

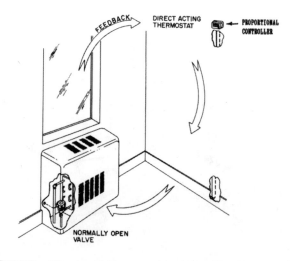

Typical Closed Loop System with proportional pneumatic controller and normally open valve with 3 to 6 PSIG spring.

As in the two-position system, the thermostat controls the valve which supplies the medium to the system, as determined by feedback through room air. However, the proportional thermostat holds the valve open to the proper position to admit just enough of the heating medium (hot water) to the coil of the terminal unit to maintain room temperature at or near the desired level.

Fig. 11-6. A closed loop system with a proportional controller.

DIFFERENTIAL

The differential is the number of units the controlled variable must change before the output signal of a two-position controller changes from minimum to maximum or vice-versa.

An open loop system as shown in Fig. 11-7, is used in certain instances, for example, when more heat is required to maintain a suitable room temperature as the outdoor temperature drops. It is possible to arrange a thermostat to measure outdoor temperature so as to cause the heat input (boiler operating temperature) of a building to increase as the outdoor temperature decreases. The outdoor thermostat cannot measure the result of heat input to the room, hence there is no feedback. Building temperature or individual room temperatures would be controlled by one or more indoor thermostats.

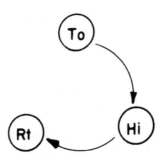

To = OUTDOOR TEMPERATURE
Hi = HEAT INPUT
Rt = ROOM TEMPERATURE

Fig. 11-7. Schematic of components of an open loop system.

All control systems are composed of one or more loops. Most will be closed loops, but some may be open loops or a combination of both. The proper combination of these elements must be applied or the "closed loop system" will not operate properly.

It is necessary to understand the difference between closed and open loops in order to appreciate the results that can be expected. There are no perfect closed loops. In actual control conditions, outside forces constantly work on the various parts to change the balance and set the loop cycle in operation to re-establish balance. If this were not the case, there would be no need for automated control.

To summarize, a closed loop system is one in which each part has an effect on the next step in the loop, and each is affected by action of the previous step. An open loop system is one in which one or more of the steps has no direct effect or action imposed on the following step or is not affected by other steps in the loop.

CONTROLLERS

A controller must be sensitive to changes in the controlled variable and respond with a precise output signal of adjustable magnitude to prevent the controlled variable from deviating too much from the set point. A controller has two main parts: the measuring or sensing element and the relay which produces the output signal. Controllers can be classified in two types: two position and proportional.

Two Position Control provides *full* delivery or *no* delivery of the power source. (Either ON or OFF).

Proportional Control influences the power source in proportion to changes in the controlled variable requirements as measured and directed by the controller. Example: A proportionally controlled device such as a damper or a valve may be partially opened or partially closed as directed by the controller.

Either type can be used. Selection is dependent on the desired results.

MEASURING ELEMENTS

The measuring element of a controller converts the status or change in a controlled variable to a useful movement or variation that will activate the relay to produce an output signal. Three types of elements commonly used in control systems measure temperature, relative humidity, and pressure.

Temperature

Temperature-measuring elements are made in various types to fit the several kinds of controllers. Wall mounted thermostats normally use bimetal or vapor-filled bellows elements. Remotely mounted or external elements have liquid, gas, or refrigerant-filled bulbs and capillaries. The capillary or connecting section on bulb types is furnished in various lengths to allow the controller to be mounted away from the measured variable.

Bimetal elements are two thin strips of dissimilar metal fused together to form a device which reliably changes its shape at a constant rate as its ambient temperature changes. Because the two metals bonded together expand and contract at vastly different rates, the measuring element, held stationary at one end, bends as the temperature varies. Thus, temperature measurement is changed into a physical movement.

Remote Bulb Elements are used when temperatures are to be measured in pipes, tanks, ducts, or relatively inaccessible locations or where a hostile environment would be detrimental to the controller. Remote bulb elements are composed of three parts: the bulb or temperature-sensitive portion; the capillary, which transmits the variable temperature signal over its length; and the diaphragm or operating head, which changes the temperature measurement into motion.

HUMIDISTATS

Humidistats are designed to measure and control the relative humidity of a building, room, or area. The measuring elements in a humidistat are usually made from from materials that attract water and respond to humidity changes by changing their size and/or shape.

PNEUMATIC TUBING

Copper tubing in sizes $1/8$", $3/16$" and $1/4$" (all O.D.-outside diameter) is used in both inaccessible and accessible locations for pneumatic tubing. Pneumatic tubing installed in or under a concrete floor, in a masonry wall, in the plaster coating of a wall, or in the hollow space of a precast concrete plank is considered inaccessible. Polyethylene tubing with barbed fittings can be used in some installations providing the tubing is installed in a concealed-accessible location.

TRAINING IN THE PNEUMATIC CONTROLS FIELD

The installation, servicing, and calibration of pneumatic control systems is a highly specialized field. The author has worked closely with JOHNSON CONTROLS (formerly *JOHNSON SERVICE CO.*) personnel in the past and suggests that pipefitters interested in receiving training in this field contact:

> JOHNSON CONTROLS INSTITUTE
> Attn: M45
> 507 E. Michigan Street
> Milwaukee, WI 53202

Portions of the text of this chapter are excerpts from various training manuals furnished by JOHNSON CONTROLS, INC. and are reproduced with permission of JOHNSON CONTROLS, INC.

CHAPTER 12

GAS PIPING

The installation, testing, and connection of gas transmission lines, gas piping, and gas-fired appliances falls within the scope of work of pipefitters and welders.

To fill a need for a single Code that would cover all facets of fuel gas piping and appliance installations, from meter set assemblies or other facilities comprising the gas service entrance to consumers' premises, a *National Fuel Gas Code* was developed by representatives of the American Gas Association (AGA), the American Society of Mechanical Engineers (ASME), and the National Fire Protection Association (NFPA). The Code was approved by the American National Standards Institute (ANSI) and is revised and updated as needed. The Code is listed as ANSI Z223.1 and NFPA 54. Copies may be obtained from ANSI and NFPA.

The *Uniform Plumbing Code* contains a section on Fuel Gas Piping.

The *Southern Building Code Congress International* has developed a Standard Gas Code.

If a question should arise regarding the installation or servicing of piping or appliances, the *local authority* (in certain Codes the *Administrative Authority*) shall have final jurisdiction. Also, the word "*shall*" is a mandatory term.

Although all gas Codes and regulations are designed to promote safety in the installation of piping and the servicing of gas-fueled equipment and appliances, and are therefore basically similar, the reader is advised to become thoroughly familiar with the Gas Code enacted in his or her area of work.

TYPES OF GASES

Manufactured Gas

There are several types of manufactured gases. The most common kind is *coke-oven* gas, a by-product of manufacturing of coke from coal. Coke-oven gas has a heating value of 525 to 570 Btu (British thermal units) per cubic foot.

Mixed Gas

Mixed gas is a mixture of natural gas with any of several types of manufactured gas.

Natural Gas

Natural gas, in its original state, is lighter than air and will escape upward in the event of leakage. If the leak is within a building where escape to the outside is impossible, the risk of fire and/or explosion is very high.

Liquid Petroleum Gas

Liquid petroleum (LP) gas is a fuel found in natural gas and is made up principally of propane or butane or a mixture of the two products. When demand is high, many utilities add liquid petroleum gas to natural gas. LP gas under moderate pressure becomes a liquid and is easily transported and stored. At normal atmospheric pressure and temperature, the liquid returns automatically to the gaseous state. The heat content of LP gas is about three times greater than that of natural gas and about six times greater than that of most manufactured gases. At atmospheric pressure, pure butane *vaporizes* or returns to gas at 32°F. Pure propane returns to gas at −44°F. Under normal conditions these are the lowest temperatures at which the liquid turns to gas. For these reasons, propane is used more in cold climates; butane is used more in mild climates.

LP gas is heavier than air and is colorless. Being heavier than air, leaking LP gas will pocket in low areas of a room, structure, or area. Explosions and fires resulting in loss of life and heavy property damage have occurred when a spark ignited pocketed LP gas. Suppliers add odors to LP gas to aid in detecting leaks.

Kinds of Piping

The kinds of piping, metallic or plastic, to be used in an installation shall be determined by the building code or regulation in effect at the job site and by job specifications drawn up by the architect/mechanical engineer and/or gas supplier. Chemical composition of gases used in an area are usually the determining factor. Black steel pipe may be acceptable in one area and not permitted in another area. Example: Whereas regular type L or type K copper tube may be acceptable in one area, tin-lined type L or type K copper tube may be required in other areas or by other suppliers. A relatively new innovation in piping material is the use of CSST, Corrugated Stainless Steel Tubing.

Gas Meters

Gas meters shall be located in readily accessible spaces and shall be placed in areas where they will not be subjected to damage. Location of gas meters is usually the responsibility of the gas supplier.

Regulators

When the gas supply pressure is more than the equipment is designed to handle, or varies beyond the design pressure limits of the equipment, a line gas pressure regulator shall be installed.

Troubleshooting Hint:

Problem: No gas supply to building.

Regulators are installed in conjunction with outside gas meters. If the venting connection of the regulator is blocked by snow, ice, or debris, the regulator may not be able to "breathe" (flex), shutting off gas supply to the building.

Low Pressure Protection

If the operation of gas utilization equipment such as a gas compressor may produce a vacuum or a dangerous reduction in gas pressure at the meter, a suitable protective device shall be installed. Such devices include, but are not limited to, mechanical, diaphragm-operated, or electrically operated low-pressure shutoff valves.

Valves

Gas valves shall be lubricated straight-way shutoff type and shall be placed in an accessible location. Valves controlling appliances shall be located within three feet of such appliances.

Combustion Air

Gas-burning appliances or devices require a supply of air for proper combustion. Improper combustion will result in sooting, causing a heavy concentration of carbon monoxide. Consult gas supplier for cubic feet of air required for proper combustion.

Testing Methods

Test pressure shall be measured with either a manometer (Fig. 12-1) or with a pressure-measuring device designed and calibrated to read, record, or indicate a pressure loss due to leakage during the test.

After test pressure has been applied, the source of pressure shall be disconnected from the piping before testing begins. Test pressure shall be no less than $1^1/_2$ times the proposed working pressure, but not less that 3 psig, irrespective of design pressure. When a manometer is used, common practice is to lift off the burner on a gas stove, then slide the free end of the manometer tubing over the burner orifice. The manometer is prepared for use by pouring water into one side of the tubing until it equalizes at the zero point of the gauge. When the water level is equalized at the zero point, connect the rubber tubing to the orifice. Open the burner valve or (if testing gas pressure) turn on the gas and observe the reading on each side of the manometer tube. Figure 12-1 (A) shows the pressure equalized at the zero point. Figure 12-1 (B) shows the water column after pressure has been applied. The reading on the side connected to the test point shows $2^1/_2$ in. below the zero point. The reading on

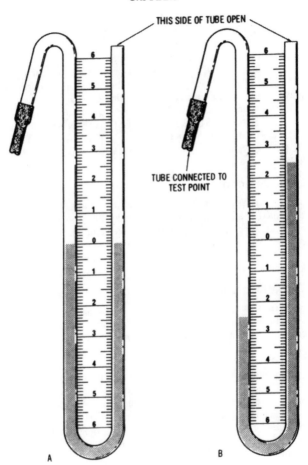

THIS SIDE OF TUBE OPEN

TUBE CONNECTED TO
TEST POINT

A

B

Fig. 12-1. Using a manometer.

the other side shows the water level at $2^1/_2$ in. above the test point. Adding the two readings together gives the pressure applied, $2^1/_2 + 2^1/_2 = 5$. If the water level equalizes during the test when using the manometer for testing, the pipe is leaking.

Test duration shall be not less than one-half hour for each 500 cubic feet of pipe volume or fraction thereof. The test duration may be reduced to ten minutes in a single family home. Any reduction in test pressure shall be deemed to indicate a leak.

CAUTION: NO SUBSTANCE OTHER THAN AIR SHALL BE INTRODUCED INTO THE GAS PIPING FOR TESTING PURPOSES. **OXYGEN** MUST **NEVER** BE USED FOR TESTING.

Leak Detection

Any reduction of test pressure indicates a leak. The leak shall be located by means of an approved combustible gas detector, soap and water, or an equivalent nonflammable solution. *Open flames, candles, matches, or any method which could provide a source of ignition shall not be used.* When the leak source is located, the leak shall be repaired and the piping system retested.

Thread Compounds (Pipe Dope)

Thread compounds (pipe dope) shall be resistant to the chemical constituents of the type gas to be used. Use only the type of thread compound recommended by the gas supplier.

Pipe Sizing

Job specifications or detailed shop drawings usually indicate the required sizes for piping to fixtures or equipment. The quantity of gas to be provided at each outlet should be determined from the manufacturer's Btu rating of the appliance to be installed. To aid in the selection of correctly sized piping, Table 12-1 lists the approximate gas consumption of the average appliance in Btu per hr.

Capacities for different sizes and lengths of pipe in cubic feet per hour for gas of 0.60 specific gravity, based on a pressure drop of 0.3" of water column, are shown in Table 12-2. In adopting a 0.3" pressure drop, due allowance for an ordinary number of fittings should be made. Table 12-2 is based on a gas of 0.60 specific gravity. If you wish to use an exact specific gravity for a particular condition, correct the values in the Table by multiplying by $\sqrt{.6/s.g.}$

To obtain the size of piping required for a certain unit, you must first determine the number of cubic feet of gas per hour consumed by the unit.

$$\text{Cu. ft. of gas/hr.} = \frac{\text{total Btu / hr required by unit}}{\text{Btu / cu. ft. of gas}}$$

Table 12-1. Approximate Gas Consumption of Typical Appliances

Appliance	Input Btu/hr (approx.)
Boiler or furnace (domestic)	100,000 to 250,000
Range (freestanding, domestic)	65,000
Built-in oven or broiler unit (domestic)	25,000
Built-in top unit (domestic)	40,000
Water heater, automatic storage (50-gal. tank)	55,000
Water heater, automatic instantaneous	
2 gal. per minute	142,800
4 gal. per minute	285,000
6 gal. per minute	428,400
Water heater, domestic, circulating or side-arm	35,000
Refrigerator	3,000
Clothes dryer, domestic	35,000

Example: It is necessary to determine the size of the piping in a gas-fired boiler installation when the burner input is 155,000 Btu/hr and the heating value per cubic foot of gas is 1000 Btu. Assume the distance from the gas meter to the boiler to be 75'.

By a substitution of values in the foregoing formula we obtain:

cu. ft. of gas/hr. = $155,000/1000 = 155$

Table 12-2. Capacity of Pipe of Different Diameters and Lengths in Cu. Ft. per Hour with Pressure Drop of 0.3" and Specific Gravity of 0.60

Length of Pipe (ft.)	Iron-pipe sizes (IPS)(inch)				
	$1/2$	$3/4$	1	$1^1/4$	$1^1/2$
15	76	172	345	750	1220
30	52	120	241	535	850
45	43	99	199	435	700
60	38	86	173	380	610
75		77	155	345	545
90		70	141	310	490
105		65	131	285	450
120			120	270	420
150			109	242	380
180			100	225	350
210			92	205	320
240				190	300
270				178	285
300				170	270
450				140	226
600				119	192

Table 12-3. Approximate Length of Thread Required for Various Sizes of Gas Pipe.

Size of pipe in inches	Approximate length of threaded portion, in inches	Approximate number of threads to be cut
$3/8$	$5/16$	10
$1/2$	$3/4$	10
$3/4$	$3/4$	10
1	$7/8$	10
$1^1/4$	1	11
$1^1/2$	1	11
2	1	11
$2^1/2$	$1^1/2$	12
3	$1^1/2$	12
4	$1^3/4$	13

From Table 12-2 it will be noted that a 1" pipe 75 ft. long will handle 155 cu. ft/hr. with 0.3" pressure drop. The determination of gas pipe sizes for other appliance units can be made in a similar manner.

Example: It is necessary to determine the pipe size of each section and outlet of a piping system with a designated pressure drop of 0.3 water column. The gas to be used has a specific gravity of 0.60 and a heating value of 1000 Btu per cubic foot.

Reference to the piping layout and a substitution of values for maximum gas demand for the various outlets, Fig. 12-2, shows that the demand for individual outlets will be as follows:

Gas demand for outlet A = 55,000/1000 = 55 cu. ft./hr.
Gas demand for outlet B = 3,000/1000 = 3 cu. ft./hr.
Gas demand for outlet C = 65,000/1000 = 65 cu. ft./hr.
Gas demand for outlet D = 116,000/1000 = 116 cu. ft./hr.

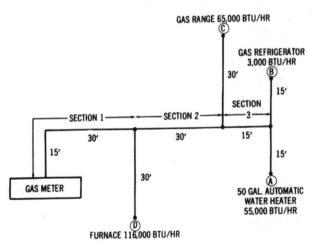

Fig. 12-2. Diagram of a gas piping layout for a typical domestic installation.

From the piping layout in Fig. 12-2 it will be noted that the length of gas pipe from the meter to the most remote outlet (A) is 105'. This is the only distance measurement used. Reference to Fig. 12-2 indicates:

> Outlet A supplying 55 cu. ft/hr. requires $1/2$" pipe.
> Outlet B supplying 3 cu. ft/hr. requires $1/2$" pipe.
> Outlet C supplying 65 cu. ft/hr. requires $3/4$" pipe.
> Outlet D supplying 116 cu. ft/hr. requires $3/4$" pipe.
> Section 3 supplying outlets A and B, or 58 cu.ft/hr., requires $1/2$" pipe.
> Section 2 supplying outlets A, B, and C, or 123 cu.ft/hr., requires $3/4$" pipe.
> Section 1 supplying outlets A, B, C, and D, or 239 cu. ft/hr., requires 1" pipe.

The determination of sizes of gas pipe for other piping layouts may be made in a similar manner.

Pipe Threads

Pipe and fittings shall comply with the American Standard for pipe threads. Pipe with damaged threads shall not be used. All gas piping shall be threaded in accordance with specifications in Table 12-3.

Branch Connections

All branches should be taken from the top or side of horizontal pipes, not from the bottom. When ceiling outlets are taken from horizontal piping, the branch should be taken from the side and carried in a horizontal direction for a distance of not less than 6". Fig. 12-3 shows the right and the wrong way to connect a drop branch to a horizontal run.

Pipe Supports

Pipe must be properly supported and not subjected to unnecessary strain. Fig. 12-4 shows a pipe supported by pipe clips or straps fastened to the building floor joists. The maximum distance

WRONG WAY   **RIGHT WAY**

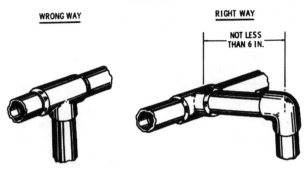

Fig. 12-3. Connection of a drop branch to a horizontal pipe.

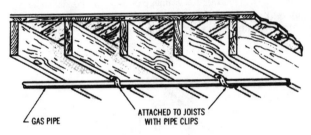

Fig. 12-4. Gas piping should be supported from the joists.

at which these supports should be spaced depends on the pipe size. The following distances between supports should never be exceeded, closer spacing being preferable.

$\frac{3}{8}$" and $\frac{1}{2}$" pipe—6'
$\frac{3}{4}$" and 1" pipe—8'
$1\frac{1}{4}$" and larger-(horizontal)—10'
$1\frac{1}{4}$" and larger-(vertical) every floor—level

Pipe shorter than the support spacing listed should also be adequately supported. Whenever a branch fitting is used, or wherever there is a change of direction of 45° or more, a support should be provided within 6" of one side of the angle fitting.

Pipe straps or iron hooks should not be used for securing pipe larger than 2". Beyond this size, when pipe is horizontal and is to be fastened to floor joists or beams, pipe hangers should be used. When the pipe is to be fastened to a wall, ring hangers and plates should be used.

When pipes run crosswise to joists or beams, do not cut the timbers deeper than one-fifth of the depth of the timbers as shown in Fig. 12-5. This cutting should be as near the support of the beam as possible, Fig. 12-6, but in no case should it be farther from a support than one-sixth of the span. When possible, pipes should be run parallel to the beams to avoid cutting and possible weakening of the beams. Horizontal piping should have some pitch as shown in Fig. 12-7 to provide for drainage of moisture condensed in the piping. A tee with a capped nipple should be provided at the lowest point to permit draining of any condensate.

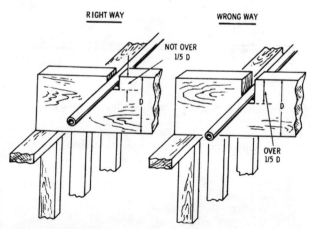

Fig. 12-5. The notches in joists should never be deeper than one-fifth the depth of the joists.

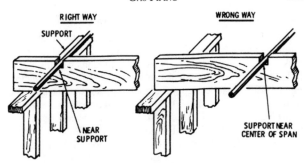

Fig. 12-6. Joists should be notched near a supporting member.

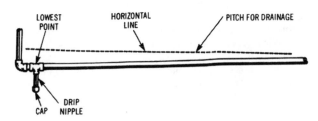

Fig. 12-7. Horizontal runs of gas pipe should have slight pitch.

Corrugated Stainless Steel Tubing (CSST)

CSST systems consist of flexible corrugated stainless steel tubing, manifolds, regulators, valves, and fittings, complete with striker plate protectors and installation instructions. The installation of Corrugated Stainless Steel Tubing (CSST) for interior fuel gas piping systems has been approved by American National Standards Institute ANSI/A.G.A/LC 1-1991. The ANSI standard is recognized in the National Fuel Gas Code of 1988, the American Gas Institute (AGA), National Fire Protection Association (NFPA), the Southern Building Code Congress Standard Gas Code, BOCA National Mechanical Code, and the CABO one and two-family Dwelling Code. The ANSI Standard applies to piping systems not exceeding 5 psig (pounds sq. in. gauge) and not exceeding a size of one inch inside diameter. CSST is not intended for installation by the general public; *installers must be trained by the manufacturer, its representatives, or other experienced organizations.*

CSST, if approved by the local administrative authority and/or by the architect or mechanical engineer, can be installed in any new or existing residential or commercial building. CSST is especially suited for the retrofit market where limited space, structural obstructions, or other difficult installation problems could exist.

Space limitations prohibit in-depth explanations of CSST systems in this book. Manufacturers of CSST systems will furnish informative literature on the installation of their products.

In the absence of a nationally adopted and enforced Fuel Gas Code, pipefitters installing fuel gas piping must conform to local regulations or codes in force at the job site.

CHAPTER 13

TUNGSTEN INERT GAS WELDING

Tungsten Inert Gas Welding (TIG) can be done on many metals, but because this book is written for pipefitters and welders, the information in this chapter pertains specifically to steel pipe and stainless steel welding. The TIG arc welding process, also called Gas Tungsten Arc Welding (GTAW), uses a non-consumable tungsten electrode to create the electric arc shown in Fig. 13-1. TIG welding can be done manually or, using a machine, automatically. The tungsten electrode is uncoated, and the arc is shielded by an inert gas or gases that flow around the tungsten tip and over the molten weld. The effect of the shielding gas to prevent contamination of the weld is shown in Fig. 13-2.

FILTER RECOMMENDATIONS (adapted from ANSI Safety Standard Z49.1)

Application	Lens Shade No.*
TIG (Tungsten Arc)	
up to 50 amps	10
50 to 150 amps	12
above 150 amps	14

*As a rule of thumb, start with a shade that is too dark to see the arc zone. Then go to a lighter shade which gives sufficient view of the arc zone without exerting a strain on your eyes.

POWER SUPPLY FOR TIG WELDING

The power supply for TIG welding may be either alternating current (ac) or direct current (dc). Certain weld characteristics obtained with each type often make one or the other better suited to a specific application. Table 13-1 shows the recommended type

211

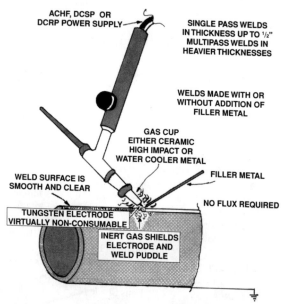

ACHF, DCSP OR
DCRP POWER SUPPLY

SINGLE PASS WELDS
IN THICKNESS UP TO 1/2"
MULTIPASS WELDS IN
HEAVIER THICKNESSES

WELDS MADE WITH OR
WITHOUT ADDITION OF
FILLER METAL

GAS CUP
EITHER CERAMIC
HIGH IMPACT OR
WATER COOLER METAL

FILLER METAL

WELD SURFACE IS
SMOOTH AND CLEAR

NO FLUX REQUIRED

TUNGSTEN ELECTRODE
VIRTUALLY NON-CONSUMABLE

INERT GAS SHIELDS
ELECTRODE AND
WELD PUDDLE

Fig. 13-1. Essentials of the TIG welding process.

of current to use for a given job. Recommended electrodes for use with both ac and dc are shown in Table 13-2.

DIRECT CURRENT WELDING

In direct current welding, the welding current circuit may be connected as either "straight polarity" or "reverse polarity." The machine connection for direct current straight polarity (DCSP) is electrode negative and work positive. The electrons flow from the electrode to the workpiece. For direct current reverse polarity welding (DCRP), the machine connection is electrode positive and workpiece negative. When straight polarity is used, the electrons exert a considerable heating effect on the workpiece; in reverse of polarity welding, where the electrode acquires this extra heat which tends to melt off the end of the electrode. Therefore, for any given welding current, DCRP requires a larger diameter electrode than does DCSP. Example: A 1/16" diameter pure tungsten electrode can handle 125 amperes of straight polarity welding current. If the polarity were reversed, this amount of current would melt off the electrode and contaminate the workpiece. Therefore, a 1/4" diameter pure tungsten electrode is needed to handle 125 amperes DCRP.

Table 13-1. Current Selection for TIG Welding

MATERIAL	ALTERNATING CURRENT* With High-Frequency Stabilization	DIRECT CURRENT STRAIGHT POLARITY	REVERSE POLARITY
Magnesium up to 1/8 in. thick	1	N.R.	2
Magnesium above 3/16 in. thick	1	N.R.	N.R.
Magnesium Castings	1	N.R.	2
Aluminum up to 3/32 in. thick	1	N.R.	2
Aluminum over 3/32 in. thick	1	N.R.	N.R.
Aluminum Castings	1	N.R.	N.R.
Stainless Steel	2	1	N.R.
Brass Alloys	2	1	N.R.
Silicon Copper	N.R.	1	N.R.
Silver	2	1	N.R.
Hastelloy Alloys	2	1	N.R.
Silver Cladding	1	N.R.	N.R.
Hard-Facing	1	1	N.R.
Cast Iron	2	1	N.R.
Low Carbon Steel, 0.015 to 0.030 in.	2**	1	N.R.
Low Carbon Steel, 0.030 to 0.125 in.	N.R.	1	N.R.
High Carbon Steel, 0.015 to 0.030 in.	2	1	N.R.
High Carbon Steel, 0.030 in. and up	2	1	N.R.
Deoxidized Copper***	N.R.	1	N.R.

KEY:
1. Excellent operation.
2. Good operation.
N.R. Not recommended.
*Where ac is recommended as a second choice, use about 25% higher current than is recommended for DCSP.
**Do not use ac on tightly jigged part.
***Use brazing flux or silicon bronze flux for 1/4-in. and thicker.

Table 13-2. Recommended Electrodes for TIG Welding

WELDING CURRENT RANGE, AMP (See note below)

Electrode Diameter in.	AC* Using Pure Tungsten Electrodes	AC* Using Thoriated Tungsten Electrodes**	DCSP Using Pure Tungsten or Thoriated Tungsten Electrodes	DCRP Using Pure Tungsten or Thoriated Tungsten Electrodes
.020	5–15	—	—	—
.040	10–60	60–80	15–80	—
1/16	50–100	100–150	70–150	10–20
3/32	100–160	160–235	150–250	15–30
1/8	150–210	225–325	250–400	25–40
5/32	200–275	300–425	400–500	40–55
3/16	250–350	400–525	500–800	55–80
1/4	325–475	500–700	800–1000	80–125

*Maximum values shown have been determined using an unbalanced wave transformer. If a balanced wave transformer is used, either reduce these values by about 30% or use the next size electrode. This is necessary because of the higher heat input to the electrode in a balanced wave setup.

**Balled electrode tip ends can best be formed and sustained at these current levels.

NOTE: The recommendations in this chart apply to genuine HELIARC tungsten electrodes.

These opposite heating effects influence not only the welding action but also the shape of the weld being made. DCSP welding will produce a narrow, deep weld; DCRP, using a larger diameter electrode, gives a wide, relatively shallow weld. When DCRP is used, the electrons and gas ions tend to remove the surface oxides and scale usually present, thus cleaning the weld area.

ALTERNATING CURRENT WELDING

When ac welding current is used, half of each ac cycle is DCSP, the other half is DCRP. Moisture, oxides, scale, etc. on the surface of the workpiece tend to prevent the flow of current in the DCRP half cycle, partially or even completely. This is called rectification. If no current were to flow in the reverse polarity direction, all cleaning action would be lost. To prevent this from happening, it is common practice to introduce into the current a high voltage, high frequency, low power additional current. This high frequency current jumps the gap between the electrode and the workpiece and pierces the oxide film, thereby forming a path for the welding current. Superimposing this high voltage, high frequency current provides the following advantages:

1. The arc may be started without touching the electrode to the workpiece.
2. Better arc stability is obtained.
3. A longer arc is possible.
4. Welding electrodes have longer life.
5. The use of wider current ranges for a specific diameter electrode is possible.

Types Of Joints

The principal basic types of joints used in TIG welding are the butt, lap, corner, edge, and tee. Regardless of the type of joint, the workpiece must be cleaned prior to welding. Manual cleaning with a wire brush or chemical solvent is usually sufficient.

Shielding Gases

The objective of TIG is to enable welds to be made without the contaminating influence of the air in the workplace. To accomplish this, shielding gas or gases flow from the torch nozzle over and around the weld arc and workpiece. The shielding gas flows at a prescribed rate measured in cubic feet per hour (CFH) or liters per hour (LPH). The selection of the proper gas or gas mixture depends on the material being welded. Gas distributors can supply this information. Argon, an inert gas, is the most commonly used shielding gas for TIG welding. Its low thermal conductivity produces a narrow, constricted arc column which allows greater variations in arc length with minimal influence on arc power and weld bead shape. For ac welding applications, argon is preferred over helium because of its superior cleaning action, arc stability, and weld appearance.

Helium is also an inert gas and produces higher arc voltages than argon for a given current setting and arc length. This produces a "hotter" arc.

Oxygen and Carbon Dioxide

These gases are chemically reactive and should not be used with TIG welding. Their high oxidation potential destroys the tungsten electrode under the heat of the arc.

Gas Flow Rate

Gas flow rate can range from a few CFH to more than 60 CFH depending on the current, the torch size, and surrounding drafts. In general, a higher current will require a larger torch and higher flow rates. In addition, gas density, or the weight of the gas relative to the air it needs to displace, has a major influence on the minimum flow rate required to effectively shield the weld. Argon is approximately 1.4 times as heavy as air and 10 times as heavy as helium. The effect of these densities relative to air is shown in Fig. 13-2. Argon, after leaving the torch nozzle, forms a blanket over the weld area, whereas helium tends to rise in turbulent fashion around the nozzle. To produce equivalent shielding effectively, the flow of helium must be 2 to $2^{1}/_{2}$ times that of argon. The same general relationship is true for mixtures of argon and helium, particularly those high in helium content. Gas flow rate must be selected with care. It is not productive or economical to use more gas than necessary to achieve good shielding. Excess gas pulls air into the welding arc, often causing porosity in the weld. A flow device that limits gas flow to the optimal range is recommended.

Gas Preflow and Postflow

Gas preflow and postflow minimize contamination of the weld zone and electrode. A preflow of shielding gases removes moisture

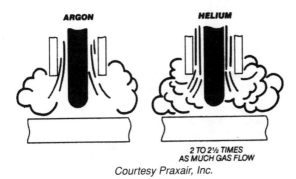

Courtesy Praxair, Inc.

Fig. 13-2. The effects of shielding gases.

which may have entered the system. Changes in room temperature can cause air to move in and out of the end of a torch while not in use, resulting in moisture condensing on the inside of the torch. When the gas is turned on, the moisture mixes with the gas as it leaves the nozzle. A preflow of shielding gas for a period of time before the arc is initiated will remove the moisture.

Postflow works in a different way. When the arc is turned off, the weld metal begins to cool, but for a few moments the weld metal remains hot enough to be contaminated by the air surrounding the just-completed weld. To prevent this, the shielding gas is allowed to flow for several seconds after the arc is extinguished. (The length of time depends on the size and temperature of the weld.) The postflow of gas also protects the hot electrode from contamination.

Backup Shielding and Trailing Shields

It is sometimes necessary to use shielding gas on the underside of a weld to prevent oxidation of the hot weld bottom. Backup shielding gas is used to purge the air from the inside of piping. This procedure prevents contamination of the backside of a weld while the pipe is being welded from the outside. In some instances, the welding may occur too fast for the shielding gas to protect the weld until it has cooled. As the arc moves on, the solidified weld metal remains hot and oxidizes. A trailing gas shield can be used to prevent this oxidation from occurring.

The Advantages of TIG Over Shielded Metal Arc Welding

In any type of welding, the best obtainable weld is one which has the same chemical, metallurgical, and physical properties as the base metal itself. To obtain such conditions, the molten weld puddle must be protected from the atmosphere during the welding operation; otherwise, atmospheric oxygen and nitrogen will combine readily with the molten weld metal and result in a weak, porous weld. In TIG welding, the weld zone is shielded from the atmosphere by an inert gas which is fed through the welding torch. Either argon or helium may be used. Argon is widely used because of its general suitability for a wide variety of metals and for the lower flow rates required. Helium provides a hotter arc, allowing 50–60% higher arc voltage for a given arc length. This extra heat is especially useful when welding heavy sections. Gas mixtures of argon and helium are used to provide the benefits of both gases. The use of non-consumable tungsten electrodes and inert shielding gases produces the highest quality welds of any open-arc welding process. Welds are bright and shiny, with no slag or spatter, and require little or no post-weld cleaning. The inert gases: argon, helium, or a combination of the two can provide 100% protection from the atmosphere, making TIG welds stronger, more ductile and more corrosion resistant than welds made with ordinary arc welding processes. Since no flux is used, corrosion due to flux entrapment cannot occur. TIG is easily used in all welding positions and provides excellent puddle control. Various combinations of dissimilar metals can be welded using the TIG process.

Starting the Arc

In ac welding, the electrode does not have to touch the workpiece to start the arc. The superimposed high frequency current jumps the gap between the welding electrode and the work, thus establishing a path for the welding current to follow. To strike an arc, first turn on the power supply and hold the torch in a horizontal position about two inches above the workpiece as shown in Fig. 13-3. Then, quickly swing the torch down toward the workpiece so that the end of the electrode is about 1/8 in. above the workpiece. The arc will then strike. This downward motion should be made rapidly to provide the maximum amount of gas protection to the weld zone. The torch position at the time the arc strikes, and the steps involved in making TIG welds are shown in Fig. 13-4.

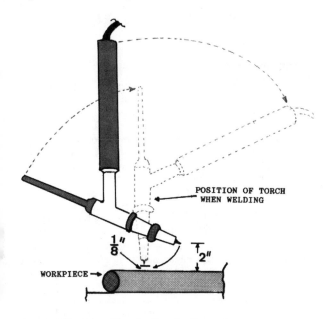

Fig. 13-3. Torch position for the starting swing.

In dc welding, the same motion is used for striking an arc. In this case, the electrode *must touch* the workpiece in order for the arc to start. As soon as the arc is struck, the electrode must be withdrawn approximately $1/8$" above the workpiece to avoid contaminating the electrode in the molten puddle. High frequency is sometimes used to start a dc arc, eliminating the need to touch the workpiece. The high frequency is automatically turned off when the arc is started.

The arc can be struck on the workpiece itself or on a heavy piece of copper or scrap steel and then carried to the starting point of the weld. Do not use a carbon block for starting the arc; the electrode would be contaminated, causing the arc to wander. When starting to weld with a hot electrode, the action must be very rapid, as the arc tends to strike before the torch is in proper position.

Stopping the Arc

To stop an arc, snap the electrode back up to the horizontal position. This motion must be made rapidly to prevent the arc from marring or damaging the weld surface or workpiece. When current density of the electrode is at a sufficiently high level, the entire end of the electrode will be in a molten state and completely covered by the arc. When too low a current density is used, only a small area of the electrode becomes molten, resulting in an unstable arc which has poor directional characteristics and is difficult for the operator to control. Too high a current density results in excessive melting of the end of the electrode. Striking an arc with a carbon pencil or on a carbon block is a primary cause of arc wandering. When the carbon touches molten tungsten, tungsten carbide is formed. Tungsten carbide has a lower melting point than pure tungsten and forms a large molten ball on the end of the electrode. This ball, in effect, reduces the current density at the electrode end, and arc wandering then occurs. The electrode can also be contaminated by touching it to the workpiece or filler rod. When electrode contamination occurs in any form, it is best to break off the contaminated end and regrind the end of the electrode.

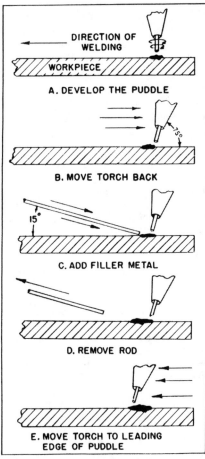

A. DEVELOP THE PUDDLE

B. MOVE TORCH BACK

C. ADD FILLER METAL

D. REMOVE ROD

E. MOVE TORCH TO LEADING
EDGE OF PUDDLE

Courtesy Praxair, Inc.

Fig. 13-4. Steps in making a TIG weld.

Arc Wandering

With the torch held stationary, the points at which the arc leaves the electrode and impinges upon the workpiece may often shift and wander without apparent reason. This is known as "arc wandering" and is generally attributed to one of the following reasons:

1. Low electrode current density
2. Carbon contamination of the electrode
3. Magnetic effects
4. Air drafts

The first two causes are distinguished by a very rapid movement of the arc from side to side, generally resulting in a zig-zag weld pattern. Magnetic effects, the third cause, generally displace the arc to one side or the other along the entire length of the weld. The fourth creates varying amounts of arc wandering, depending upon the amount of air draft present.

TIG Pipe Welding

1. Root pass, rolled joint. This will be a standard vee joint with a 37° bevel on each side, a $1/16$ in. nose and a $1/16$–$3/32$ in. spacing.

   After the joint has been tacked (four tacks 90° apart) the arc is struck on the side and carried down to the bottom of the joint. Filler metal is added until the puddle bridges over the gap. After the puddle bridges the joint, the arc is held until the puddle flattens out and becomes wedge shaped, straight across the front and rounded at the rear. When this takes place, the puddle has fully penetrated the joint.

2. Filler pass. Weave or zig-zag beads can be used on low alloy steel in the rolled position. Stringer beads laid parallel to the joint are used for stainless steel pipe. Weave or zig-zag welds can be used on low alloy steel pipe. Stringer beads are required for stainless steel heavy wall pipe in all positions and for low alloy pipe in the vertical-fixed position.

3. Finish Pass. The final pass should be about $1/16"$ above the surface of the pipe.

 The chief advantage of TIG welding for the fabrication of piping systems is that the welds are smooth, fully penetrated, and free from obstructions or crevices on the inside. TIG welds are stronger and more resistant to corrosion than welds made by any other process. Another advantage is that in today's economy production costs are very important, and TIG welding results in savings in labor costs by eliminating the labor involved in removal of slag and spatters.

Safety Precautions

 Keep your head out of the fumes while welding.
 Provide sufficient ventilation.
 Welding inside tanks, boilers, or confined spaces requires the use of an air-supplied hood or a hose mask.
 Ground all electrical equipment and the workpiece, Fig. 13-5.

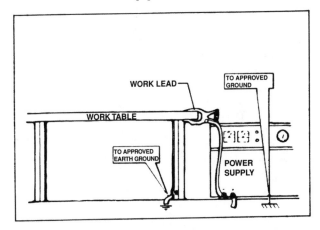

Fig. 13-5 Ground all electrical equipment and the workpiece.

These are only a few of the many safety precautions welders and pipefitters should take. For complete information regarding safety on the job, follow the recommendations found in:

American National Standard Z49.1, "Safety in Welding and Cutting" available from the American Welding Society, P.O. Box 351040, Miami, FL 33135, and also the National Electrical Code, NFPA No. 70, available from the National Fire Protection Association, Batterymarch Park, Quincy, MA 02269.

CHAPTER 14

TROUBLE-SHOOTING TIPS FOR ARC WELDING

Trouble	Cause	Remedy
Welder runs but soon stops	Wrong relay heaters.	Renewal part recommendations.
	Welder overloaded.	Considerable overload can be carried only for a short time.
	Duty cycle too high.	Do not operate continually at overload currents.
	Leads too long or too narrow in cross section.	Should be large enough to carry welding current without excessive voltage drop.
	Power circuit single-phased.	Check for one dead fuse or line.
	Ambient temperature too high.	Operate at reduced loads where temperature exceeds 100° F.
	Ventilation blocked.	Check air inlet and exhaust openings.
Welding arc is loud and spatters excessively	Current setting too high.	Check setting and output with ammeter.
	Polarity wrong.	Check polarity, try reversing, or an electrode of opposite polarity.

Trouble	Cause	Remedy
Welding arc sluggish	Current too low.	Check output, and current recommended for electrode being used.
	Poor connections.	Check all electrode-holder, cable, and ground-cable connections. Strap iron is poor ground return.
	Cable too long or too small.	Check cable voltage drop and change cable.
Touching set gives shock	Frame not grounded.	Ground solidly.
Generator control falls to vary current	Any part of field circuit may be short circuited or open circuited.	Find faulty contact and repair.
Welder starts but will not deliver welding current	Wrong direction of rotation.	See INITIAL STARTING.
	Brushes worn or missing.	Check that all brushes bear on commutator with sufficient tension.
	Brush connections loose.	Tighten.
	Open field circuit.	Check connection to rheostat, resistor, and auxiliary brush studs.
	Series field and armature circuit open.	Check with test lamp or bell ringer.
	Wrong driving speed.	Check nameplate against speed of motor or belt drive.
	Dirt, grounding field coils.	Clean and reinsulate.
	Welding terminal shorted.	Electrode holder or cable grounded.
Welder generating but current falls off when welding	Electrode or ground connection loose.	Clean and tighten all connections.
	Poor ground.	Check ground-return circuit.

(continues)

Trouble	Cause	Remedy
	Brushes worn off.	Replace with recommended grade. Sand to fit. Blow out carbon dust.
	Weak brush spring pressure.	Replace or readjust brush springs.
	Brush not properly fitted.	Sand brushes to fit.
	Brushes in backwards.	Reverse.
	Wrong brushes used.	Renewal part recommendations.
	Brush pigtails damaged.	Replace brushes.
	Rough or dirty commutator.	Turn down or clean commutator.
	Motor connection single-phased.	Check all connections.
Welder will not start (Starter not operating)	Power circuit dead.	Check voltage.
	Broken power lead.	Repair.
	Wrong supply voltage.	Check nameplate against supply.
	Open power switches.	Close.
	Blown fuses.	Replace.
	Overload relay tripped.	Let set cool. Remove cause of overloading.
	Open circuit to starter button.	Repair.
	Defective operating coil.	Replace.
	Mechanical obstruction in contactor.	Remove.
Welder will not start (Starter operating)	Wrong motor connections.	Check connection diagram.
	Wrong supply voltage.	Check nameplate against supply.

Trouble	Cause	Remedy
Welder will not start (Starter operating) (*cont.*)	Rotor stuck.	Try turning by hand.
	Power circuit single-phased.	Replace fuse; repair open line.
	Starter single-phased.	Check contact of starter tips.
	Poor motor connection.	Tighten.
	Open circuit in windings.	Repair.
Starter operates and blows fuse	Fuse too small.	Should be two to three times rated motor current.
	Short circuit in motor connections.	Check starter and motor leads for insulation from ground and from each other.

CHAPTER 15

PIPE WELDERS' DEFINITIONS

alloy steel — A steel which owes its distinctive properties to elements other than carbon.

area of a circle — The measurement of the surface within a circle. To find the area of a circle, multiply the product of the radius times the radius by Pi (3.142). Commonly written A = r².

braze weld or brazing — A process of joining metals using a nonferrous filler metal or alloy, the melting point of which is higher than 800°F but lower than that of the metals to be joined.

butt weld — A circumferential weld in pipe fusing the abutting pipe walls completely from inside wall to outside wall.

carbon steel — A steel which owes its distinctive properties chiefly to the various percentages of carbon (as distinguished from the other elements) which it contains.

circumference of a circle — The measurement around the perimeter of a circle. To find the circumference, multiply Pi (3.142) by the diameter. (Commonly written as d).

coefficient of expansion — A number indicating the degree of expansion or contraction of a substance.

The coefficient of expansion is not constant and varies with changes in temperature. For linear expansion it is expressed as the change in length of one unit of length of a substance having one degree rise in temperature. A Table of Expansion (see Appendix, Table A-11) is generally used to determine expansion or contraction within a piping system.

corrosion — The gradual destruction or alteration of a metal or alloy caused by direct chemical attack or by electrochemical reaction.

creep — The plastic flow of pipe within a system; the permanent set in metal caused by stresses at high temperatures. Generally associated with a time rate of deformation.

diameter of a circle — A straight line drawn through the center of a circle from one extreme edge to the other. Equal to twice the radius.

ductility — The property of elongation above the elastic limit but under the tensile strength.

A measure of ductility is the percentage of elongation of the fractured piece over its original length.

elastic limit — The greatest stress which a material can withstand without a permanent deformation after release of the stress.

erosion — The gradual destruction of metal or other material by the abrasive action of liquids, gases, solids, or mixtures thereof.

radius of a circle — A straight line drawn from the center to the extreme edge of a circle.

socket fitting — A fitting used to join pipe in which the pipe is inserted into the fitting. A fillet weld is then made around the edge of the fitting and the outside wall of the pipe at the junction of the pipe and fitting.

soldering — A method of joining metals using fusable alloys, usually tin and lead, having melting points under 700°F.

strain — Change of shape or size of a body produced by the action of a stress.

stress — The intensity of the internal, distributed forces which resist a change in the form of a body. When external forces act on a body they are resisted by reactions within the body which are termed stresses.

A Tensile Stress is one that resists a force tending to pull a body apart.

A Compressive Stress is one that resists a force tending to crush a body.

A Shearing Stress is one that resists a force tending to make one layer of a body slide across another layer.

A Torsional Stress is one that resists forces tending to twist a body.

tensile strength — The maximum tensile stress which a material will develop. The tensile strength is usually considered to be the load in pounds per square inch at which a test specimen ruptures.

turbulence — Any deviation from parallel flow in a pipe due to rough inner walls, obstructions, or directional changes.

velocity — Time rate of motion in a given direction and sense, usually expressed in feet per second.

volume of a pipe — The measurement of the space within the walls of the pipe. To find the volume of a pipe, multiply the length (or height) of the pipe by the product of the inside radius times the inside radius by Pi (3.142). Commonly written as $V = h r^2$.

welding — A process of joining metals by heating until they are fused together or by heating and applying pressure until there is a plastic joining action. Filler metal may or may not be used.

yield strength — The stress at which a material exhibits a specified limiting permanent set.

CHAPTER 16

DEFINITIONS OF HEATING AND AIR CONDITIONING TERMS

The following terms apply to heating and air conditioning, both in work and in designing of systems.

Absolute Humidity — The weight of water vapor in grains actually contained in one cubic foot of the mixture of air and moisture.

Absolute Pressure — The actual pressure above zero; the atmospheric pressure added to the gauge pressure. It is also expressed as a unit pressure such as lbs. per sq. in. (psig).

Absolute Temperature — The temperature of a substance measured above absolute zero. To express a temperature as absolute temperature, add 460° to the reading of a Fahrenheit thermometer or 273° to the reading of a Centigrade thermometer.

Absolute Zero — The temperature (– 460°F. approx.) at which all molecular motion of a substance ceases and at which the substance contains no heat. Scientists are working to achieve a temperature of *absolute zero,* but this goal may be unattainable.

Air — An elastic gas; a mixture of oxygen and nitrogen with slight traces of other gases. It may contain moisture known as humidity. Dry Air weighs 0.075 lbs. per cu. ft. Air expands or contracts approximately 1/490th of its volume for each degree rise or fall in temperature from 32°F.

Air Change — The number of times in an hour that the air in an area is changed, either by mechanical means or by infiltration of outside air.

Air Cleaner — A device designed for removal of airborne impurities like dust, dirt, smoke, and fumes.

Air Conditioning — The simultaneous control of temperature, humidity, air cleaning, and air movement and distribution within an area.

Air Infiltration — The leakage of air into a building through cracks, crevices, doors, windows, and other openings.

Air Vent — A device designed to purge air from radiation or piping in a steam or hot-water heating system.

Atmospheric Pressure — The weight upon a given area of a column of air one square inch in cross section extending upward to the outer limits of the atmosphere. Atmospheric pressure at sea level is approximately 14.7 pounds per square inch. Atmospheric pressure is less at the top of a mountain, more below sea level.

Boiler — In heating terms, a closed vessel in which water is heated to produce steam or hot water for heating purposes.

Boiler Heating Surface — The heat-transmitting surfaces of a boiler in contact with hot water or steam on one side and hot gases or fire on the other side.

Boiler Horsepower — The equivalent evaporation of 34.5 lbs. of water per hour at 212°F. to steam at 212°F. This equals a heat output of 33,475 British thermal units (Btu) per hour.

British Thermal Unit — The quantity of heat required to raise the temperature of one pound of water one degree F.

Bucket Trap (inverted) — A float trap with a float open at the bottom. When the air or steam in the bucket has been replaced by condensate, the bucket loses its bouyancy and sinks. As it sinks, it opens a valve to permit condensate to be returned to the boiler.

Bucket Trap (open) — An open bucket trap is open at the top. Water surrounding the bucket keeps it floating, but as condensate drains into the bucket, the bucket sinks, opening an outlet allowing steam pressure to force the condensate out of the trap.

Calorie (small) — The quantity of heat required to raise one (1) gram of water one (1) degree C.

Calorie (large) — The quantity of water required to raise one (1) kilogram of water one (1) degree C.

Celsius — A thermometer scale at which the freezing point of water is 0° and its boiling point is 100°.

Chimney Effect — The tendency of air in a duct to rise or fall as density of the air varies with temperature.

Coefficient of Heat Transmission (U) — The amount of heat (Btu) transmitted in one hour, from *air to air,* per square foot of wall, floor, roof, or ceiling for a difference in temperature of one (1) degree F. *between the air on the inside and outside of the wall, floor, roof, or ceiling.*

Comfort Line — The effective temperature at which the largest percentage of adults feel comfortable.

Comfort Zone — The range of effective temperatures at which the largest percentage of adults feel comfortable.

Condensate — In steam heating, the water formed as steam cools. One pound of condensate per hour is equal to approximately 4 sq. ft. of steam heating surface (240 Btu per hour per sq. ft.).

Conductivity (thermal)-k — The amount of heat (Btu) transmitted in one hour through one square foot of a homogenous material one inch thick for a difference in temperature of one degree F. between the two surfaces of the material.

Convection — The transmission of heat by the circulation, either natural or forced, of a liquid or a gas such as air. Natural convection is caused by the difference of weight or density of a hotter or colder fluid or air.

Convector — A concealed or enclosed heating or cooling unit transferring heat or cooled air by the process of convection.

Convertor — A vessel or device for heating water with steam without mixing the two. The process is called *indirect* heating.

Cooling Leg — A length of uninsulated pipe through which condensate flows to a trap and which has enough cooling surface to allow the condensate to dissipate adequate heat to prevent "flashing" when the trap opens. (see Flash).

Degree Day — A unit which is the difference between 65°F and the daily average temperature when the latter is below 65°F. The "degree days" in any one day is equal to the number of degrees F that the average temperature for that day is below 65°F.

Dew-Point Temperature — The air temperature corresponding to saturation (100% relative humidity) for a given moisture content. It is the lowest temperature at which air can retain the moisture it contains.

Direct-Indirect Heating Unit — A heating unit which is partially enclosed, the enclosed portion being used to heat air which enters from outside the room.

Direct Return System (Hot Water) — A two-pipe hot water heating system in which the water, after it has passed through a heating unit, is returned to the boiler along a direct path so that the total distance traveled by the water from each radiator is the shortest distance feasible. Each circuit in a system will vary in length.

Domestic Hot Water — Hot water used for purposes other than house heating, such as bathing, cooking, laundering, and dishwashing.

Down-Feed One-Pipe Riser (Steam) — A pipe which carries steam downward to the heating units and into which condensate from the heating units drains.

Down-Feed System (Steam) — A steam heating system in which the supply mains are above the level of the heating systems which they serve.

Dry-Bulb Temperature — The air temperature as determined by an ordinary thermometer.

Dry Return (Steam) — A return pipe in a steam heating system which carries both condensation water and air.

Dry Saturated Steam — Saturated steam which carries no water in suspension. (See Saturated Steam).

Equivalent Direct Radiation (EDR) — The amount of heating surface which will give off 240 Btu per hour when filled with a heating medium at 215°F. and surrounded by air at 70°F. (The equivalent square feet of heating surface may have no direct relation to the actual surface area.)

Fahrenheit — A thermometer scale at which the freezing point of water is 32° and its boiling point is 212° above zero.

Flash — The instant passing into steam of water at a high temperature when the pressure it is under is reduced so that its temperature is above that of its boiling point for the reduced pressure. For example: If hot condensate is discharged by a trap into a low pressure return or into the atmosphere, a percentage of the water will be immediately transformed into steam. Another term for flash is re-evaporation.

Float and Thermostatic Trap — A float trap with a thermostatic element for permitting the escape of air into the return line.

Float Trap — A steam trap which is operated by a float. When enough condensate has drained into the trap body, the float is lifted, opening a port, thereby permitting the condensate to flow into the return and letting the float drop. When the float has been sufficiently lowered, the port is closed again. Temperature does not affect the operation of a float trap.

Gauge Pressure — The pressure above that of the atmosphere and indicated on the gauge. It is expressed as pounds per square inch gauge (psig).

Head — Pressure expressed in feet of water. One foot of water column exerts a pressure of .43 lbs.

Heat — That form of energy into which all other forms may be changed. Heat always flows from a body of higher temperature to a body of lower temperature. (See: Latent Heat, Sensible Heat, Specific Heat, Total Heat, Heat of the Liquid.)

Heat of the Liquid — The heat (Btu) contained in a liquid by virtue of its temperature. The heat of the liquid for water is zero at 32°F. and increases approximately 1 Btu for every degree rise in temperature.

Heat Unit — In the foot-pound-second system, the British thermal unit (Btu); in the centimeter-gram-second system, the calorie (cal.).

Heating Medium — A substance such as water, steam, or air used to convey heat from any source of heat to the heating units from which the heat is dissipated.

Heating Surface — The exterior surface of a heating unit.

Heating unit — Any device which transmits heat from a heating system or source to an area and its occupants.

Horsepower — A unit to indicate the time rate of doing work equal to 550 ft-lb. per second or 33,000 ft-lb. per minute. One horsepower equals 2545 Btu per hour or 746 watts.

Humidistat — An instrument which controls the relative humidity of an area.

Humidity — Water vapor present in air.

Insulation (Thermal) — Material with a high resistance to heat flow.

Latent Heat of Evaporation — The heat (Btu per pound) necessary to change one pound of liquid into vapor without raising its temperature. In round numbers this is equal to 960 Btu per pound of water.

Latent Heat of Fusion — The heat necessary to melt one pound of a solid without raising the temperature of the resulting liquid. The latent heat of fusion of water (melting one pound of ice) is 144 Btu.

Mechanical Equivalent of Heat — The mechanical energy equivalent to 1 Btu which is equal to 778 ft-lb.

Mil-Inch — One one-thousandth of an inch (0.001).

One-Pipe Supply Riser (Steam) — A pipe which carries steam to a heating unit and which also carries the condensation from the heating unit. In an up-feed riser, steam travels upward and condensate travels downward while in a down-feed system, both steam and condensate travel down.

One-Pipe System (Hot Water) — A hot-water heating system in which one pipe serves both as a supply main and a return main. The heating units have separate supply and return connections but both connect back to one main.

One-Pipe System (Steam) — A steam heating system in which both steam and condensate flow in the same main. Each heating unit has only one connection which serves as both steam supply and condensate return connection.

Overhead System — Any steam or hot water system in which the supply main is above the heating units. With a steam system, the return must be below the heating units; with a water system, the return may be above the heating units.

Pressure — Force per unit area such as lb. per sq. in. Usually refers to unit static gauge pressure. (See Static, Velocity, and Total Gauge and Absolute Pressures).

Pressure Reducing Valve — A device for changing the pressure of a gas or liquid from a higher pressure to a lower pressure.

Radiant Heating — A heating system in which the heating is by radiation only.

Radiation — The transmission of heat in a straight line through space.

Radiator — A heating unit exposed to view and located within the room to be heated. A radiator transfers heat by radiation to objects "it can see" and by conduction to the surrounding air which, in turn, is circulated by natural convection.

Reducing Valve — (See Pressure Reducing Valve.)

Re-evaporation — (See Flash.)

Refrigeration, Ton of — (See Ton of Refrigeration.)

Relative Humidity — The amount of moisture in a given quantity of air compared with the maximum amount of moisture the same quantity of air could hold at the same temperature. It is expressed as a percentage.

Return Mains — The pipes which return the heating medium from the heating units to the source of heat supply.

Reverse-Return System (Hot Water) — A two-pipe hot water heating system in which the water from the several heating units is returned along paths arranged so that all radiator circuits of the system are practically of even length.

Sensible Heat — Heat which only increases the temperature of objects as opposed to latent heat.

Specific Heat — In the foot-pound-second system, the amount of heat required to raise one pound of a substance one degree F. In the centimeter-gram-second system, the amount of heat (cal.) required to raise one gram of a substance one degree centigrade. The specific heat of water is 1.

Square Foot of Heating Surface — (See Equivalent Direct Radiation).

Static Pressure — The pressure which tends to burst a pipe. It is used to overcome the frictional resistance to flow through a pipe. It is expressed as a unit pressure and may be either in absolute or gauge pressure. It is frequently expressed in feet of water column or (in the case of pipe friction) in mil-inches of water column per foot of pipe.

Steam — Water in the vapor phase. The vapor formed when water has been heated to its boiling point, corresponding to the pressure it is under. (See also Dry Saturated Steam, Wet Saturated Steam, Superheated Steam.)

Steam Heating System — A heating system in which the units give up their heat to the area by condensing the steam furnished to them by the boiler or other source.

Steam Trap — A device for allowing the passage of condensate and air but preventing the passage of steam. (See Thermostatic, Float, Bucket trap.)

Superheated Steam — Steam heated above the temperature corresponding to its pressure.

Supply Mains — The pipes through which the heating medium flows from the boiler or source of supply to the run-outs and risers leading to the heating units.

Thermostat — An instrument which responds to changes in temperature and which directly or indirectly controls area temperature.

Thermostatic Trap — A steam trap which opens by a drop in temperature such as when cold condensate or air reaches it and closes when steam reaches it. The temperature-sensitive element is usually a sealed bellows or series of diaphragm chambers containing a small quantity of volatile liquid.

Ton of Refrigeration — The heat which must be extracted from one ton, (2000 lbs.) of water at 32°F. to change it into ice at 32°F. in 24 hrs. It is equal to 288,000 Btu/24 hrs, 12,000 Btu/hr. or 200 Btu/minute.

Total Heat — The latent heat of vaporization added to the heat of the liquid with which it is in contact.

Total Pressure — The sum of the static and velocity pressures. It is also used as the total static pressure over an entire area (i.e., the unit pressure multiplied by the area on which it acts).

Trap — (See Steam Trap, Thermostatic Trap, Float Trap, Bucket Trap.)

Two-Pipe System (Steam or Water) — A heating system in which one pipe is used for the supply main and another for the return main. The essential feature of a two-pipe hot water system is that each heating unit receives a direct supply of the heating medium which cannot have served a preceding heating unit.

Unit Pressure — Pressure per unit area as lbs. per sq. in. (psig).

Up-Feed System (Hot Water or Steam) — A heating system in which the supply mains are below the level of the heating units which they serve.

Vacuum Heating System (Steam) — A one- or two-pipe heating system equipped with the necessary accessory apparatus to permit the pressure in the system to go below atmospheric.

Vapor — Any substance in the gaseous state.

Vapor Heating System (Steam) — A two-pipe heating system which operates under pressure at or near atmospheric and which returns the condensate to the boiler or receiver by gravity.

Velocity Pressure — The pressure used to create the velocity of flow in a pipe.

Vent Valve (Steam) — A device for permitting air to be forced out of a heating unit or pipe which then closes against water and steam.

Vent Valve (Water) — A device that permits air to be pushed out of a pipe or heating unit but that closes against water.

Wet Bulb Temperature — The lowest temperature which a water-wetted body will attain when exposed to an air current.

Wet Return (Steam) — That part of a return main of a steam heating system which is completely filled with condensation water.

Wet Saturated Steam — Saturated steam containing some water particles in suspension.

CHAPTER 17

GLOSSARY OF TERMS RELATING TO PLASTIC PIPING

Adhesive — a substance capable of holding materials together by surface attachment.

Adhesive, solvent — an adhesive having a volatile organic liquid as a vehicle. See Solvent Cement.

Aging, n. — (1) the effect on materials of exposure to an environment for an interval of time. (2) the process of exposing materials to an environment for an interval of time.

Antioxidant — a compounding ingredient added to a plastic composition to retard possible degradation from contact with oxygen (air), particularly in processing at or exposure to high temperatures.

Artificial weathering — the exposure of plastics to cyclic laboratory conditions involving changes in temperature, relative humidity, and ultraviolet radiant energy, with or without direct water spray, in an attempt to produce changes in the material similar to those observed after longterm continuous outdoor exposure.

> *NOTE:* The laboratory exposure conditions are usually intensified beyond those encountered in actual outdoor exposure in an attempt to achieve an accelerated effect. This definition does not involve exposure to special conditions such as ozone, salt spray, industrial gases, etc.

Bell end — the enlarged portion of a pipe that resembles the socket portion of a fitting and that is intended to be used to make a joint by inserting a piece of pipe into it. Joining may be accomplished by solvent cements, adhesives, or mechanical techniques.

Beam loading — the application of a load to a pipe between two points of support, usually expressed in pounds, and the distance between the centers of the supports.

Burst strength — the internal pressure required to break a pipe or fitting. This pressure will vary with the rate of build-up of the pressure and the time during which the pressure is held.

Cement — See adhesive and solvent cement.

Chemical resistance — (1) the effect of specific chemicals on the properties of plastic piping with respect to concentration, temperature and time of exposure. (2) the ability of a specific plastic pipe to render service for a useful period in the transport of a specific chemical at a specified concentration and temperature.

Chlorinated Poly (Vinyl Chloride) Plastics — plastics made by combining chlorinated poly (vinyl chloride) with colorants, fillers, plasticizers, stabilizers, lubricants, and other compounding ingredients.

Cleaner — medium strength organic solvent such as methylethyl ketone to remove foreign matter from pipe and fitting joint surfaces.

Compound — the intimate admixture of a polymer or polymers with other ingredients such as fillers, softeners, plastics, catalysts, pigments, dyes, curing agents, stabilizers, antioxidants, etc.

Copolymer — See Polymer.

Creep, n. — the time-dependent part of strain resulting from stress, that is, the dimensional change caused by the application of load over and above the elastic deformation and with respect to time.

Cv — See Flow Coefficient.

Deflection temperature — the temperature at which a specimen will deflect a given distance at a given load under prescribed conditions of test. See ASTM D648. Formerly called heat distortion.

Degradation, n. — a deleterious change in the chemical structure of a plastic. See also Deterioration.

Deterioration — a permanent change in the physical properties of a plastic evidenced by impairment of these properties.

> *NOTE:* Burst strength, fiber stress, hoop stress, hydrostatic design stress, long-term hydrostatic strength, hydrostatic strength (quick) long-term burst, ISO equation, pressure, pressure rating, quick burst, service factor, strength stress, and sustained pressure test are related terms.

Elasticity — that property of plastic materials by virtue of which they tend to recover their original size and shape after deformation.

> *NOTE:* If the strain is proportional to the applied stress, the material is said to exhibit Hookean or ideal elasticity.

Elastomer — a material which at room temperature can be stretched repeatedly to at least twice its original length and which, immediately upon release of the stress, will return with force to its approximate original length.

Elevated temperature testing — tests on plastic pipe above 23°C (73°F).

Environmental stress cracking — cracks that develop when the material is subjected to stress in the presence of specific chemicals.

Extrusion — a method whereby heated or unheated plastic forced through a shaping orifice becomes one continuously formed piece.

> *NOTE:* This method is commonly used to manufacture thermoplastic pipe.

Failure, adhesive — rupture of an adhesive bond, such that the plane of separation appears to be at the adhesive-adherend interface.

Fiber stress — the unit stress, usually in pounds per square inch (psig), in a piece of material that is subjected to an external load.

Filler — a relatively inert material added to a plastic to modify its strength, permanence, working properties, or other qualities, or to lower costs.

Flow Coefficient or Cv — valve coefficient of flow representing the flow rate of water in gallons per minute which will produce a 1 psig pressure drop through the valve.

Full Port Valve — one in which the resistance to flow, in the open position, is equal to an equivalent length of pipe.

Fungi resistance — the ability of plastic pipe to withstand fungi growth and/or their metabolic products under normal conditions of service or laboratory tests simulating such conditions.

Heat joining — making a pipe joint by heating the edges of the parts to be joined so that they fuse and become essentially one piece with or without the addition of additional material.

Hoop stress — the tensile stress, usually in pounds per square inch (psig), in the circumferential orientation in the wall of the pipe when the pipe contains a gas or liquid under pressure.

Hydrostatic design stress — the estimated maximum tensile stress in the wall of the pipe in the circumferential orientation due to internal hydrostatic pressure that can be applied continuously with a high degree of certainty that failure of the pipe will not occur.

Hydrostatic strength (quick) — the hoop stress calculated by means of the ISO equation at which the pipe breaks due to an internal pressure build-up, usually within 60 to 90 seconds.

Impact, Izod — a specific type of impact test made with a pendulum type machine. The specimens are molded or extruded with a machined notch in the center. See ASTM D256.

ISO equation — an equation showing the interrelations between stress, pressure and dimensions in pipe, namely

$$S = [P (ID + t)/2t] \text{ or } [P (OD - t)/2t]$$
where S = stress
P = pressure
ID = average inside diameter
OD = average outside diameter
t = minimum wall thickness
Reference: ISO R161-1960 Pipes of Plastic Materials for the Transport of Fluids (Outside Diameters and Nominal Pressures) Part I, Metric Series.

Joint — the location at which two pieces of pipe or a pipe and a fitting are connected. The joint may be made by an adhesive, a solvent-cement, or a mechanical device such as threads or a ring seal.

Long-term burst — the internal pressure at which a pipe or fitting will break due to a constant internal pressure held for 100,000 hours (11.43 years).

Long-term hydrostatic strength — the estimated tensile stress in the wall of the pipe in the circumferential orientation (hoop stress) that when applied continuously will cause failure of the pipe at 100,000 hours (11.43 years). These strengths are usually obtained by extrapolation of log-log regression equations or plots.

Molding, injection — a method of forming plastic objects from granular or powdered plastics by the fusing of plastic in a chamber with heat and pressure and then forcing part of the mass into a cooler chamber where it solidifies.

NOTE: — This method is commonly used to manufacture thermoplastic fittings.

Outdoor exposure — plastic pipe placed in service or stored so that it is not protected from the elements of normal weather conditions, i.e., the sun's rays, rain, air and wind. Exposure to industrial and waste gases, chemicals, engine exhausts, etc., are not considered normal "outdoor exposure."

Permanence — the property of a plastic which describes its resistance to appreciable changes in characteristics with time and environment.

Plastic, n. — a material that contains as an essential ingredient an organic polymeric substance of large molecular weight, is solid in its finished state, and, at some stage in its manufacture or in its processing into finished articles, can be shaped by flow.

Plastic pipe — a hollow cylinder of a plastic material in which the wall thicknesses are usually small when compared to the diameter and in which the inside and outside walls are essentially concentric. See plastic tubing.

Plastic tubing, n. — a particular size of plastics pipe in which the outside diameter is essentially the same as that of copper tubing. See plastic pipe.

Polypropylene, n. — a polymer prepared by the polymerization of propylene as the sole monomer. See Polypropylene plastics.

Polypropylene plastics — plastics based on polymers made with propylene as essentially the sole monomer.

Poly (vinyl chloride) — a polymer prepared by the polymerization of vinyl chloride as the sole monomer.

Poly (vinyl chloride) plastics — plastics made by combining poly (vinyl chloride) with colorants, fillers, plasticizers, stabilizers, lubricants, other polymers, and other compounding ingredients. Not all of these modifiers are used in pipe compounds.

Pressure — when expressed with reference to pipe, the force per unit area exerted by the medium in the pipe.

Pressure rating — the estimated maximum pressure that the medium in the pipe can exert continuously with a high degree of certainty that failure of the pipe will not occur.

Primer — strong organic solvent, preferably tetrahydrofuran, used to dissolve and soften the joint surfaces in preparation for and prior to the application of solvent cement. Primer is usually tinted purple.

PVDF — a crystaline, high molecular weight polymer of vinylidene fluoride, containing 59 percent fluorine by weight.

Quick burst — the internal pressure required to burst a pipe or fitting due to an internal pressure build-up, usually within 60 to 90 seconds.

Schedule — a pipe size system (outside diameters and wall thicknesses) originated by the iron pipe industry.

Self-extinguishing — the ability of a plastic to resist burning when the source of heat or flame that ignited it is removed.

Service factor — a factor which is used to reduce a strength value to obtain an engineering design stress. The factor may vary depending on the service conditions, the hazard, the length of service desired, and the properties of the pipe.

Solvent cement — in the plastic piping field, a solvent adhesive that contains a solvent that dissolves or softens the surfaces being bonded so that the bonded assembly becomes essentially one piece of the same type of plastic.

Solvent cementing — making a pipe joint with a solvent cement. See Solvent cement.

Stress — when expressed with reference to pipe, the force per unit area in the wall of the pipe in the circumferential orientation due to internal hydrostatic pressure.

Sustained pressure test — a constant internal pressure test for 1000 hours.

Thermoplastic — a plastic which is thermoplastic in behavior. Capable of being repeatedly softened by increase of temperature and hardened by decrease of temperature.

Throttling valve — a valve that is used for control of flow rate.

Union — a device placed in a pipeline to facilitate disassembly of the system.

Vinyl chloride plastics — plastics based on polymers of vinyl chloride or copolymers of vinyl chloride with other monomers, the vinyl chloride being in greatest amount by mass.

Weld- or Knit-line — a mark on a molded plastic part formed by the union of two or more streams of plastic flowing together.

Appendix

MISCELLANEOUS INFORMATION

251

Table A.1. ABBREVIATIONS APPLYING TO THE PIPING TRADES

Abbreviations conform to the practice of the American Standard Abbreviations for Scientific and Engineering Terms, ASA Z10.1.

abs	Absolute
AGA	American Gas Association
AISI	American Iron and Steel Institute
Amer Std	American Standard
API	American Petroleum Institute
ASA	American Standards Association
ASHVE	American Society of Heating and Ventilating Engineers
ASME	American Society of Mechanical Engineers
ASTM	American Society for Testing Materials
avg	Average
AWWA	American Water Works Association
B or Bé	Baumé
B & S	Bell and spigot or Brown & Sharpe (gauge)
bbl	Barrel
Btu	British thermal unit(s)
C	Centigrade
Cat	Catalogue
cfm	Cubic feet per minute
cfs	Cubic feet per second
CI	Cast iron
CS	Cast steel (not recommended for abbreviation)
Comp	Companion
cu ft	Cubic feet
cu in.	Cubic inch(es)

continues

Table A.1. ABBREVIATIONS APPLYING TO THE PIPING TRADES *(Cont'd)*

C to F	Center to face
deg or °	Degree(s)
°C	Degrees Centigrade
°F	Degrees Fahrenheit
diam	Diameter
dwg	Drawing
ex-hy	Extra-heavy
F&D	Faced and drilled
F	Fahrenheit
F to F	Face to face
Fig	Figure
flg	Flange or flanges
flgd	Flanged
g	Gage or gauge
gal	Gallon
galv	Galvanized
gpm or gal per min	Gallons per minute
hex	Hexagonal
hg	Mercury
hr	Hour
IBBM	Iron body bronze (or brass) mounted
ID	Inside diameter
IPS	Iron pipe size (now obsolete—see NPS)
kw	Kilowatt(s)
lb	Pound(s)
max	Maximum
Mfr	Manufacturer

Table A.1. ABBREVIATIONS APPLYING TO THE PIPING TRADES *(Cont'd)*

MI	Malleable iron
min	Minimum
MSS	Manufacturers Standardization Society (of Valve and Fittings Industry)
mtd	Mounted
NEWWA	New England Water Works Association
NPS	Nominal pipe size (formerly IPS for iron pipe size)
OD	Outside diameter
OS&Y	Outside screw and yoke
OWG	Oil, water, gas (see WOG)
psig	Pounds per square inch, gage
red	Reducing
scd	Screwed
sched or sch	Schedule
sec	Second
SF	Semifinished
Spec	Specification
sq	Square
SSP	Steam service pressure
SSU	Seconds Saybolt Universal
Std	Standard
Trans	Transactions
WOG	Water, oil, gas (see OWG)
wt	Weight
WWP	Working water pressure
XS	Extra strong
XXS	Double extra strong

Table A.2. PROPERTIES OF AIR

		DRY AIR		
Temperature °F	Weight per Cu. Ft. of Dry Air in Lbs.	Ratio to Volume at 70°F	Btu Absorbed per Cu. Ft. of Air per °F	Cu. Ft. of Air Raised 1°F by 1 Btu
0	.08636	.8680	.02080	48.08
10	.08453	.8867	.02039	49.05
20	.08276	.9057	.01998	50.05
30	.08107	.9246	.01957	51.10
40	.07945	.9434	.01919	52.11
50	.07788	.9624	.01881	53.17
60	.07640	.9811	.01846	54.18
70	.07495	1.0000	.01812	55.19
80	.07356	1.0190	.01779	56.21
90	.07222	1.0380	.01747	57.25
100	.04093	1.0570	.01716	58.28
110	.06968	1.0756	.01687	59.28
120	.06848	1.0945	.01659	60.28
130	.06732	1.1133	.01631	61.32
140	.06620	1.1320	.01605	62.31
150	.06510	1.1512	.01578	63.37
160	.06406	1.1700	.01554	64.35
170	.06304	1.1890	.01530	65.36
180	.06205	1.2080	.01506	66.40

Table A.2. PROPERTIES OF AIR *(Cont'd)*

| | | SATURATED AIR | | |
Temperature °F	Vapor Press. Inches of Mercury	Weight of Vapor per Cu. Ft. in Lbs.	Btu Absorbed per Cu. Ft. of Air per °F	Cu. Ft. of Air Raised 1°F by 1 Btu
0	0.0383	.000069	.02082	48.04
10	0.0631	.000111	.02039	49.50
20	0.1030	.000177	.01998	50.05
30	0.1640	.000276	.01955	51.15
40	0.2477	.000409	.01921	52.06
50	0.3625	.000587	.01883	53.11
60	0.5220	.000829	.01852	54.00
70	0.7390	.001152	.01811	55.22
80	1.0290	.001576	.01788	55.93
90	1.4170	.002132	.01763	56.72
100	1.9260	.002848	.01737	57.57
110	2.5890	.003763	.01716	58.27
120	3.4380	.004914	.01696	58.96
130	4.5200	.006357	.01681	59.50
140	5.8800	.008140	.01669	59.92
150	7.5700	.010310	.01663	60.14
160	9.6500	.012956	.01664	60.10
170	12.2000	.016140	.01671	59.85
180	15.2900	.019940	.01682	59.45

Table A.3. BARLOW'S FORMULA

Barlow's Formula is used to find the relationship between internal fluid pressure and stress in the pipe wall. It is simple to use and is conservative; the results are safe. Barlow's Formula is sometimes known as the "outside diameter" formula because it utilizes the outside diameter of the pipe. Bursting tests on commercial steel pipe of the commonly used thicknesses have shown that Barlow's Formula predicts the pressure at which the pipe will rupture with an accuracy well within the limits of uniformity of commercial pipe thickness.

$$P = \frac{2 \times t \times S}{D}$$

where:

P = internal units pressure, psig

S = unit stress, psig

D = outside diameter of pipe, in.

t = wall thickness, in.

Table A.4. BOILING POINTS OF WATER AT VARIOUS PRESSURES

V = Vacuum in inches of Mercury

B.P. = Boiling point of water

Gauge = psig (pounds per square in. gauge)

Vacuum	Boiling Point	Vacuum	Boiling Point	Gauge	Boiling Point
29	79.62	14	181.82	0	212
28	99.93	13	184.61	1	215.6
27	114.22	12	187.21	2	218.5
26	124.77	11	189.75	3	
25	133.22	10	192.19	4	224.4
24	140.31	9	194.50	5	
23	146.45	8	196.73	6	229.8
22	151.87	7	198.87	7	
21	156.75	6	200.96	8	234.8
20	161.19	5	202.25	9	
19	165.24	4	204.85	10	239.4
18	169.00	3	206.70	15	249.8
17	172.51	2	208.50	25	266.8
16	175.80	1	210.25	50	297.7
15	178.91			75	320.1
				100	337.9
				125	352.9
				200	387.90

Table A.5. PIPE CLAMPS

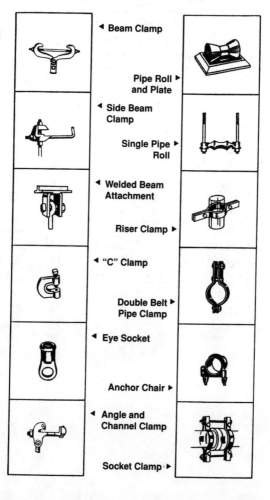

◄ Beam Clamp

Pipe Roll ►
and Plate

◄ Side Beam
Clamp

Single Pipe ►
Roll

◄ Welded Beam
Attachment

Riser Clamp ►

◄ "C" Clamp

Double Belt ►
Pipe Clamp

◄ Eye Socket

Anchor Chair ►

◄ Angle and
Channel Clamp

Socket Clamp ►

Table A.6. COLORS AND APPROXIMATE
TEMPERATURE FOR CARBON STEEL

Color	Temperature
Black Red	990°F
Dark Blood Red	1050°F
Dark Cherry Red	1175°F
Medium Cherry Red	1250°F
Full Cherry Red	1375°F
Light Cherry, Scaling	1550°F
Salmon, Free Scaling	1650°F
Light Salmon	1725°F
Yellow	1825°F
Light Yellow	1975°F
White	2220°F

Table A.7. CRANE SIGNALS FOR MOBILE AND OVERHEAD CRANES

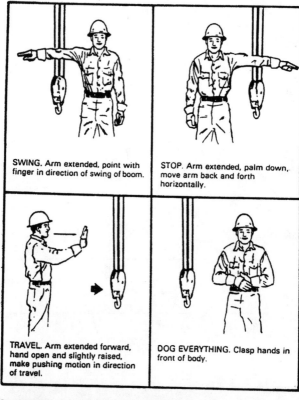

SWING. Arm extended, point with finger in direction of swing of boom.

STOP. Arm extended, palm down, move arm back and forth horizontally.

TRAVEL. Arm extended forward, hand open and slightly raised, make pushing motion in direction of travel.

DOG EVERYTHING. Clasp hands in front of body.

Standard hand signals for controlling crane operations.

ANSI/ASME B30.5 Mobile and Locomotive Cranes. (Reproduced by permission of the American Society of Automotive Engineers)

Table A.7. CRANE SIGNALS FOR MOBILE AND OVERHEAD CRANES *(Cont'd)*

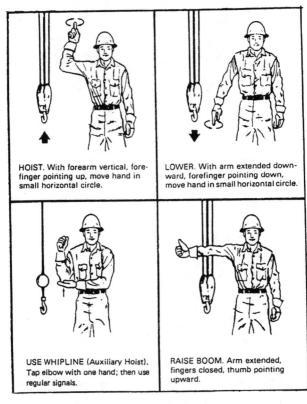

HOIST. With forearm vertical, fore-finger pointing up, move hand in small horizontal circle.

LOWER. With arm extended down-ward, forefinger pointing down, move hand in small horizontal circle.

USE WHIPLINE (Auxiliary Hoist). Tap elbow with one hand; then use regular signals.

RAISE BOOM. Arm extended, fingers closed, thumb pointing upward.

Standard hand signals for controlling crane operations *(Cont'd)*.

continues

Table A.7. CRANE SIGNALS FOR MOBILE AND OVERHEAD CRANES *(Cont'd)*

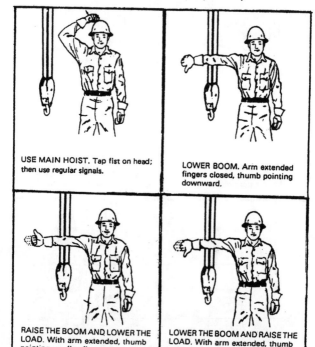

USE MAIN HOIST. Tap fist on head; then use regular signals.

LOWER BOOM. Arm extended fingers closed, thumb pointing downward.

RAISE THE BOOM AND LOWER THE LOAD. With arm extended, thumb pointing up, flex fingers in and out as long as load movement is desired.

LOWER THE BOOM AND RAISE THE LOAD. With arm extended, thumb pointing down, flex fingers in and out as long as load movement is desired.

Standard hand signals for controlling crane operations *(Cont'd)*.

Table A.7. CRANE SIGNALS FOR MOBILE AND OVERHEAD CRANES *(Cont'd)*

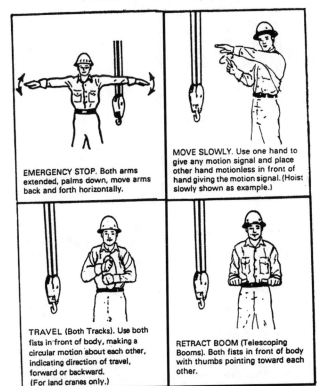

EMERGENCY STOP. Both arms extended, palms down, move arms back and forth horizontally.

MOVE SLOWLY. Use one hand to give any motion signal and place other hand motionless in front of hand giving the motion signal. (Hoist slowly shown as example.)

TRAVEL (Both Tracks). Use both fists in front of body, making a circular motion about each other, indicating direction of travel, forward or backward. (For land cranes only.)

RETRACT BOOM (Telescoping Booms). Both fists in front of body with thumbs pointing toward each other.

Standard hand signals for controlling crane operations *(Cont'd).*

continues

Table A.7. CRANE SIGNALS FOR MOBILE AND OVERHEAD CRANES *(Cont'd)*

TRAVEL. (One Track) Lock the track on side indicated by raised fist. Travel opposite track in direction indicated by circular motion of other fist, rotated vertically in front of body. (For land cranes only.)

EXTEND BOOM (Telescoping Booms). Both fists in front of body with thumbs pointing outward.

EXTEND BOOM (Telescoping Boom). One Hand Signal. One fist in front of chest with thumb tapping chest.

RETRACT BOOM (Telescoping Boom). One Hand Signal. One fist in front of chest, thumb pointing outward and heel of fist tapping chest.

Standard hand signals for controlling crane operations *(Cont'd)*.

Table A.7. CRANE SIGNALS FOR MOBILE AND OVERHEAD CRANES (Cont'd)

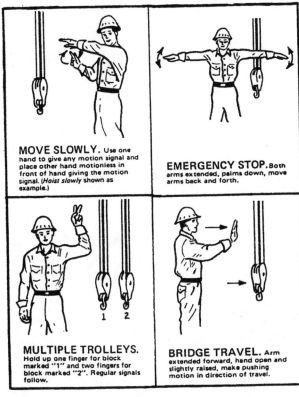

MOVE SLOWLY. Use one hand to give any motion signal and place other hand motionless in front of hand giving the motion signal. (*Hoist slowly* shown as example.)

EMERGENCY STOP. Both arms extended, palms down, move arms back and forth.

MULTIPLE TROLLEYS. Hold up one finger for block marked "1" and two fingers for block marked "2". Regular signals follow.

BRIDGE TRAVEL. Arm extended forward, hand open and slightly raised, make pushing motion in direction of travel.

Standard hand signals for controlling crane operations (Cont'd).

ANSI/ASME B30.11 Cab-operated Monorails and Underhung Cranes. (*Reproduced by permission of the American Society of Automotive Engineers*)

Table A.8. CUBIC FEET TO U.S. GALLONS, IMPERIAL GALLONS, AND LITERS

Cu. Ft.	U.S. Gals.	Imp. Gals.	Liters
1	7.5	6.2	28
2	15.0	12.5	57
3	22.4	18.7	85
4	29.9	24.9	113
5	37.4	31.2	142
6	44.9	37.4	170
7	52.4	43.6	198
8	59.8	49.9	227
9	67.3	56.1	255
10	74.8	62.4	283
20	149.6	124.7	566
30	224.4	187.1	849
40	299.2	249.4	1,133
50	374.0	311.8	1,416
60	448.8	374.1	1,699
70	523.6	436.5	1,982
80	598.4	498.8	2,265
90	673.2	561.2	2,548
100	748.0	623.6	2,832
200	1,496.1	1,247.1	5,663
300	2,244.2	1,870.7	8,495
400	2,992.2	2,494.2	11,327
500	3,740.3	3,117.8	14,158
600	4,488.3	3,741.3	16,990
700	5,236.4	4,364.9	19,822

Table A.8. CUBIC FEET TO U.S. GALLONS, IMPERIAL GALLONS, AND LITERS *(Cont'd)*

Cu. Ft.	U.S. Gals.	Imp. Gals.	Liters
800	5,984.4	4,988.4	22,653
900	6,732.5	5,612.0	25,485
1,000	7,480.5	6,235.5	28,317
2,000	14,961.0	12,471.0	56,633
3,000	22,441.6	18,706.5	84,950
4,000	29,922.1	24,942.0	113,266
5,000	37,402.6	31,177.5	141,583
6,000	44,883.1	37,413.0	169,900
7,000	52,363.6	43,648.5	198,216
8,000	59,844.1	49,884.0	226,533
9,000	67,324.7	56,119.5	254,849
10,000	74,805.2	62,355.0	283,166
20,000	149,610.4	124,710.	566,332
30,000	224,415.6	187,065.	849,498
40,000	299,220.8	249,420.	1,132,664
50,000	374,025.9	311,775.	1,415,830
60,000	448,831.1	374,130.	1,698,996
70,000	523,636.3	436,485.	1,982,162
80,000	598,441.5	498,840.	2,265,328
90,000	673,246.7	561,195.	2,548,494
100,000	748,051.9	623,550.	2,831,660

Table A.9. DISCHARGE—FIRE STREAMS

Gallons of Water per Minute Discharged from One or More 1⅛-inch Smooth-Bore Hose Nozzles Playing Simultaneously and Attached to 200 Feet of Best Quality Rubber-Lined Hose, with Pressure at the Hose-Connection Varying Between 10 and 150 Pounds.

No. 1 ⅛" Hose Nozzles	PRESSURE AT HOSE-CONNECTION IN POUNDS PER SQUARE INCH														
	Feeble Streams			Ordinary Fire Streams						Unusually Strong Streams					
	10	20	30	40	50	60	70	80	90	100	110	120	130	140	150
1	89	129	157	182	202	222	240	257	272	287	300	314	326	339	351
2	178	258	314	364	404	444	480	514	544	574	600	628	652	678	702
3	267	387	471	546	606	666	720	771	816	861	900	942	978	1017	1053
4	356	516	628	728	808	888	960	1028	1088	1148	1200	1256	1304	1356	1404
5	445	645	785	910	1010	1110	1200	1285	1360	1435	1500	1570	1630	1695	1755
6	534	774	942	1092	1212	1332	1440	1542	1632	1722	1800	1884	1956	2034	2106
7	623	903	1099	1274	1414	1554	1680	1799	1904	2009	2100	2198	2282	2373	2457
8	712	1032	1256	1456	1616	1776	1920	2056	2176	2296	2400	2512	2608	2712	2808
9	801	1161	1413	1638	1818	1998	2160	2313	2448	2583	2700	2826	2934	3051	3159
10	890	1290	1570	1820	2020	2220	2400	2570	2720	2870	3000	3140	3260	3390	3510
11	979	1419	1727	2002	2222	2442	2640	2827	2992	3157	3300	3454	3586	3729	3861
12	1068	1548	1884	2184	2424	2664	2880	3084	3264	3444	3600	3768	3912	4068	4212
13	1157	1677	2041	2366	2626	2886	3120	3341	3536	3731	3900	4082	4238	4407	4563
14	1246	1806	2198	2548	2828	3108	3360	3598	3808	4018	4200	4396	4564	4746	4914
15	1335	1935	2355	2730	3030	3330	3600	3855	4080	4305	4500	4710	4890	5085	5265

Table A.10. DRILL SIZES FOR PIPE TAPS

SIZE OF TAP IN INCHES	NUMBER OF THREADS PER INCH	DIAM. OF DRILL	SIZE OF TAP IN INCHES	NUMBER OF THREADS PER INCH	DIAM. OF DRILL
1/8	27	11/32	2	11 1/2	2 3/16
1/4	18	7/16	2 1/2	8	2 9/16
3/8	18	37/64	3	8	3 3/16
1/2	14	23/32	3 1/2	8	3 11/16
3/4	14	59/64	4	8	4 3/16
1	11 1/2	1 5/32	4 1/2	8	4 3/4
1 1/4	11 1/2	1 1/2	5	8	5 5/16
1 1/2	11 1/2	1 49/64	6	8	6 5/16

Table A.11. TOTAL THERMAL EXPANSION OF PIPING MATERIAL IN INCHES PER 100 FT. ABOVE 32°F.

TEMP. °F	CARBON AND CARBON MOLY STEEL	CAST IRON	COPPER	BRASS AND BRONZE	WROUGHT IRON
32	0	0	0	0	0
100	0.5	0.5	0.8	0.8	0.5
150	0.8	0.8	1.4	1.4	0.9
200	1.2	1.2	2.0	2.0	1.3
250	1.7	1.5	2.7	2.6	1.7
300	2.0	1.9	3.3	3.2	2.2
350	2.5	2.3	4.0	3.9	2.6
400	2.9	2.7	4.7	4.6	3.1
450	3.4	3.1	5.3	5.2	3.6
500	3.8	3.5	6.0	5.9	4.1
550	4.3	3.9	6.7	6.5	4.6

continues

Table A.11. TOTAL THERMAL EXPANSION OF PIPING MATERIAL IN INCHES PER 100 FT. ABOVE 32°F. *(Cont'd)*

TEMP. °F	CARBON AND CARBON MOLY STEEL	CAST IRON	COPPER	BRASS AND BRONZE	WROUGHT IRON
600	4.8	4.4	7.4	7.2	5.2
650	5.3	4.8	8.2	7.9	5.6
700	5.9	5.3	9.0	8.5	6.1
750	6.4	5.8	-	-	6.7
800	7.0	6.3	-	-	7.2
850	7.4	-	-	-	-
900	8.0	-	-	-	-
950	8.5	-	-	-	-
1000	9.1	-	-	-	-

Table A.12. TYPICAL BTU VALUES OF FUELS

ASTM RANK SOLIDS	BTU VALUES PER LB.
Anthracite Class I	11,230
Bituminous Class II Group 1	14,100
Bituminous Class II Group 3	13,080
Sub-Bituminous Class III Group 1	10,810
Sub-Bituminous Class III Group 2	9,670

LIQUIDS	BTU VALUES PER GAL.
Fuel Oil No. 1	138,870
Fuel Oil No. 2	143,390
Fuel Oil No. 4	144,130
Fuel Oil No. 5	142,720
Fuel Oil No. 6	137,275

Table A.12. TYPICAL BTU VALUES OF FUELS *(Cont'd)*

GASES	BTU VALUES PER CU. FT.
Natural Gas	935 to 1132
Producers Gas	163
Illuminating Gas	534
Mixed (Coke oven and water gas)	545

Table A.13. PIPE HANGERS

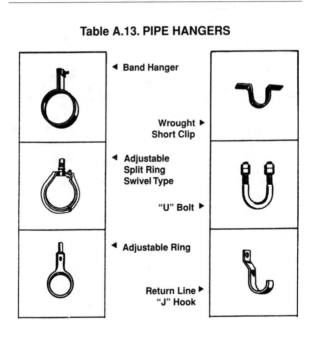

◄ Band Hanger

Wrought ► Short Clip

◄ Adjustable Split Ring Swivel Type

"U" Bolt ►

◄ Adjustable Ring

Return Line ► "J" Hook

continues

Table A.13. PIPE HANGERS *(Cont'd)*

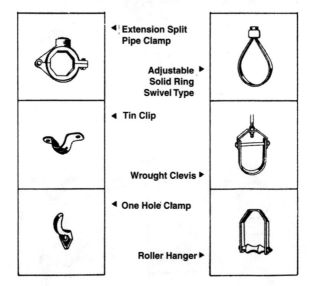

◄ Extension Split
Pipe Clamp

Adjustable ►
Solid Ring
Swivel Type

◄ Tin Clip

Wrought Clevis ►

◄ One Hole Clamp

Roller Hanger ►

Table A.14. HEADS OF WATER IN FEET WITH EQUIVALENT PRESSURES

Feet Head	Pounds per Square Inch	Feet Head	Pounds per Square Inch
1	.43	100	43.31
2	.87	110	47.64
3	1.30	120	51.97
4	1.73	130	56.30
5	2.17	140	60.63
6	2.60	150	64.96
7	3.03	160	69.29
8	3.46	170	73.63
9	3.90	180	77.96
10	4.33	200	86.62
15	6.50	250	108.27
20	8.66	300	129.93
25	10.83	350	151.58
30	12.99	400	173.24
40	17.32	500	216.55
50	21.65	600	259.85
60	25.99	700	303.16
70	30.32	800	346.47
80	34.65	900	389.78
90	38.98	1000	433.00

Note: One foot of water at 62° Fahrenheit equals .433 pound pressure per square inch. To find the pressure per square inch for any feet head not given in the table above, multiply the feet head by .433.

Table A.15. HEATING AND VENTILATING SYMBOLS

HIGH PRESSURE STEAM SUPPLY PIPE	
LOW PRESSURE STEAM SUPPLY PIPE	
HOW WATER PIPE-FLOW (OR WET RETURN)	
RETURN PIPE-STEAM OR WATER (OR DRY RETURN)	
AIR VENT LINE	

FLANGES		COLUMN RADIATOR (PLAN)	
SCREWED UNION		COLUMN RADIATOR (ELEVATION)	
ELBOW		COLUMN RADIATOR (END VIEW)	
ELBOW LOOKING UP			
ELBOW LOOKING DOWN		WALL RADIATOR (PLAN)	
TEE		WALL RADIATOR (ELEVATION)	
TEE LOOKING UP			
TEE LOOKING DOWN		WALL RADIATOR (END VIEW)	
GATE VALVE			
GLOBE VALVE		PIPE COIL (PLAN)	
ANGLE VALVE		PIPE COIL (ELEVATION)	
ANGLE VALVE (STEM PERPENDICULAR)			
LOCK SHEILD VALVE		PIPE COIL (END VIEW)	
CHECK VALVE		INDIRECT RADIATOR (PLAN)	
REDUCING VALVE		INDIRECT RADIATOR (ELEVATION)	
DIAPHRAGM VALVE		INDIRECT RADIATOR (END VIEW)	
DIAPHRAGM VALVE (STEM PERPENDICULAR)		SUPPLY DUCT (SECTION)	
THERMOSTAT		EXHAUST DUCT (SECTION)	
RADIATOR TRAP (ELEVATION)		BUTTERFLY DAMPER (PLAN OR ELEVATION)	
RADIATOR TRAP (PLAN)		BUTTERFLY DAMPER (ELEVATION OR PLAN)	
EXPANSION JOINT		DEFLECTING DAMPER (SQUARE PIPE)	
AIR SUPPLY OUTLET			
EXHAUST OUTLET		VANES	

Table A.16. HEAT LOSSES FROM HORIZONTAL BARE STEEL PIPE

(BTU per hour per linear foot at 70°F room temperature.)

NOM. PIPE SIZE	HOT WATER (180°F)	STEAM 5 PSIG (20 PSIA)
1/2	60	96
3/4	73	118
1	90	144
1 1/4	112	179
1 1/2	126	202
2	155	248
2 1/2	185	296
3	221	355
3 1/2	244	401
4	279	448

Table A.17. Conversion Table
(Inches to Centimeters to Millimeters)

in.	cm.	mm.
1.00	2.54	25.40
2.00	5.08	50.80
3.00	7.62	76.20
4.00	10.16	101.60
5.00	12.70	127.00
6.00	15.24	152.40
7.00	17.78	177.80
8.00	20.32	203.20
9.00	22.86	228.60
10.00	25.40	254.00
20.00	50.80	508.00
30.00	76.20	762.00
36.00	91.40	914.00
40.00	101.60	1016.00
50.00	127.00	1270.00
60.00	152.40	1524.00
70.00	177.80	1778.00
80.00	203.20	2032.00
90.00	228.60	2286.00
100.00	254.00	2540.00

Table A.18. LAYING OUT ANGLES WITH A SIX-FOOT FOLDING RULE. MARKINGS ON WHICH END OF RULE IS PLACED TO FORM VARIOUS ANGLES

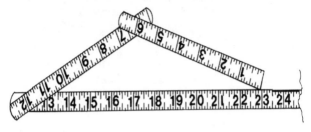

ANGLE	SET ON
5 5/8	24"
11 1/4	23 15/16"
22 1/2	23 3/4"
45	23"
60	22 1/4

Table A-36 shows a 45° angle formed with a six-foot folding rule. To form this angle unfold the rule and place the tip on 23 inches. The first two sections of the rule form a 45° angle. A six foot folding rule is marked in 1/16 in. increments therefore the numbers in the chart are to the nearest 1/16 in.

Table A.19. SYMBOLS, WEIGHTS, AND MELTING POINTS OF METALS

Material	Chemical Symbol	Weight, in Pounds per Cubic Foot	Weight, in Pounds per Cubic Inch	Melting Point, Degrees Fahrenheit
Aluminum	Al	.093	160	1218
Antimony	Sb	.2422	418	1150
Brass	..	.303	524	1800
Bronze	..	.320	552	1700
Chromium	Cr	.2348	406	2740
Copper	Cu	.323	450	2450
Gold	Au	.6975	1205	1975
Iron (cast)	Fe	.260	450	2450
Iron (wrought)	Fe	.2834	490	2900
Lead	Pb	.4105	710	620
Manganese	Mn	.2679	463	2200
Mercury	Hg	.491	849	−39.5
Molybdenum	Mo	.309	534	4500
Monel	..	.318	550	2480
Platinum	Pt	.818	1413	3200
Steel (mild)	Fe	.2816	490	2600
Steel (stainless)	..	.277	484	2750
Tin	Sn	.265	459	450
Titanium	Ti	.1278	221	3360
Zinc	Zn	.258	446	787

Table A.20. CONVERSION TABLE
(METRIC TO ENGLISH)

METRIC	ENGLISH

MEASURES OF LENGTH

METRIC	ENGLISH
1 Kilometer	0.621 Mile
1000 Meters	3281. Feet
1 Meter	1.094 Yards
100 Centimeters	3.28 Feet
1000 Millimeters	39.37 Inches
1 Centimeter	0.0328 Feet
10 Millimeters	0.394 Inches
1 Millimeter	0.0394 Inches

MEASURES OF SURFACE

METRIC	ENGLISH
1 Sq. Kilometer	0.386 Sq. Mile
1,000,000 Sq. Meters	247.1 Acres
	1,195,985 Sq. Yards
1 Sq. Meter	1.196 Sq. Yards
10,000 Sq. Centimeters	10.76 Sq. Feet
	1550. Sq. Inches
1 Sq. Centimeter	0.155 Sq. Inch
100 Sq. Millimeters	0.0011 Sq. Feet
1 Sq. Millimeter	0.00155 Sq. Inch

continues

Table A.20. CONVERSION TABLE (METRIC TO ENGLISH) *(Cont'd)*

MEASURES OF VOLUME AND CAPACITY

1 Cu. Meter	1.308 Cu. Yards
1000 Liters	35.31 Cu. Feet
1,000,000 Cu. Centimeters	61023.4 Cu. Inches
1 Liter	0.264 Gallons (U.S.)
1,000 Cu. Centimeters	0.220 Gallons (Imperial)
	1.057 Quarts (U.S.)
	0.880 Quarts (Imperial)
1 Cu. Centimeter	0.061 Cu. Inches
1000 Cu. Millimeters	

MEASURES OF WEIGHT

1 Kilogram	0.0011 Ton (2000 Lbs.)
1000 Gram	2.205 Pounds (Av.)
1 Gram	0.0022 Pounds (Av.)
	0.035 Ounces (Av.)
	15.43 Grains

Table A.21. MULTIPLIERS THAT ARE USEFUL TO THE TRADE

To Change	To	Multiply by
Inches	Feet	0.0833
Inches	Millimeters	25.4
Feet	Inches	12.
Feet	Yards	3.
Yards	Feet	.03333
Square inches	Square feet	0.00694

Table A.21. MULTIPLIERS THAT ARE USEFUL TO THE TRADE *(Cont'd)*

To Change	To	Multiply by
Square feet	Square inches	144.
Square feet	Square yards	0.11111
Square yards	Square feet	9.
Cubic inches	Cubic feet	0.00058
Cubic feet	Cubic inches	1728.
Cubic feet	Cubic yards	0.03703
Cubic yards	Cubic feet	27.
Cubic inches	Gallons	0.00433
Cubic feet	Gallons	7.48
Gallons	Cubic inches	231.
Gallons	Cubic feet	0.1337
Gallons	Pounds of water	8.33
Pounds of water	Gallons	0.12004
Ounces	Pounds	0.0625
Pounds	Ounces	16.
Inches of water	Lbs. per sq. inch	0.0361
Inches of water	Inches of mercury	0.0735
Inches of water	Ounces per sq. inch	0.578
Inches of water	Lbs. per sq. foot	5.2
Inches of mercury	Inches of water	13.6
Inches of mercury	Feet of water	1.1333
Inches of mercury	Lbs. per sq. inch	0.4914
Ounces per sq. inch	Inches of mercury	0.127
Ounces per sq. inch	Inches of water	1.733
Pounds per sq. inch	Inches of water	27.72
Pounds per sq. inch	Feet of water	2.310
Pounds per sq. inch	Inches of mercury	2.04
Pounds per sq. inch	Atmospheres	0.0681
Feet of water	Lbs. per sq. inch	0.434
Feet of water	Lbs. per sq. foot	62.5
Feet of water	Inches of mercury	0.8824

continues

Table A.21. MULTIPLIERS THAT ARE USEFUL TO THE TRADE *(Cont'd)*

To Change	To	Multiply by
Atmospheres	Lbs. per sq. inch	14.696
Atmospheres	Inches of mercury	29.92
Atmospheres	Feet of water	34.
Long tons	Pounds	2240.
Short tons	Pounds	2000.
Short tons	Long tons	0.89285

Table A.22. EXPANSION OF PIPES IN INCHES PER 100 FEET

Temperature °F	Cast Iron	Wrought Iron	Steel	Brass or Copper
0	0.00	0.00	0.00	0.00
50	0.36	0.40	0.38	0.57
100	0.72	0.79	0.76	1.14
125	0.88	0.97	0.92	1.40
150	1.10	1.21	1.15	1.75
175	1.28	1.41	1.34	2.04
200	1.50	1.65	1.57	2.38
225	1.70	1.87	1.78	2.70
250	1.90	2.09	1.99	3.02
275	2.15	2.36	2.26	3.42
300	2.35	2.58	2.47	3.74
325	2.60	2.86	2.73	4.13
350	2.80	3.08	2.94	4.45

Table A.23. PROPERTIES OF SATURATED STEAM (APPROX.)

Absolute Pressure (POUNDS PER SQ INCH)	*Gage Reading at Sea Level* (PRESSURE-POUNDS PER SQ INCH)	*Temp. °F.*	*Heat in Water B.T.U. per Lb.*	*Latent Heat in Steam (Vaporization) B.T.U. per Lb.*	*Volume of 1 Lb. Steam Cu. Ft.*	*Wgt. of Water Lbs. per Cu. Ft.*
0.18	29.7	32	0.0	1076	3306	62.4
0.50	29.4	59	27.0	1061	1248	62.3
1.0	28.9	79	47.0	1049	653	62.2
2.0	28	101	69	1037	341	62.0
4.0	26	125	93	1023	179	61.7
6.0	24	141	109	1014	120	61.4
8.0	22	152	120	1007	93	61.1
10.0	20	161	129	1002	75	60.9
12.0	18	169	137	997	63	60.8
14.0	16	176	144	993	55	60.6
16.0	14	182	150	989	48	60.5
18.0	12	187	155	986	43	60.4
20.0	10	192	160	983	39	60.3
22.0	8	197	165	980	36	60.2
24.0	6	201	169	977	33	60.1
26.0	4	205	173	975	31	60.0
28.0	2	209	177	972	29	59.9
29.0	1	210	178	971	28	59.9
30.0	0	212	180	970	27	59.8
14.7	0	212	180	970	27	59.8
15.7	1	216	184	968	25	59.8
16.7	2	219	187	966	24	59.7
17.7	3	222	190	964	22	59.6

continues

Table A.23. PROPERTIES OF SATURATED STEAM (APPROX.) *(Cont'd)*

Absolute Pressure	Gage Reading at Sea Level	Temp. °F.	Heat in Water B.T.U. per Lb.	Latent Heat in Steam (Vaporization) B.T.U. per Lb.	Volume of 1 Lb. Steam Cu. Ft.	Wgt. of Water Lbs. per Cu. Ft.
18.7	4	225	193	962	21	59.5
19.7	5	227	195	960	20	59.4
20.7	6	230	198	958	19	59.4
21.7	7	232	200	957	19	59.3
22.7	8	235	203	955	18	59.2
23.7	9	237	205	954	17	59.2
25	10	240	208	952	16	59.2
30	15	250	219	945	14	58.8
35	20	259	228	939	12	58.5
40	25	267	236	934	10	58.3
45	30	274	243	929	9	58.1
50	35	281	250	924	8	57.9
55	40	287	256	920	8	57.7
60	45	293	262	915	7	57.5
65	50	298	268	912	7	57.4
70	55	303	273	908	6	57.2
75	60	308	277	905	6	57.0
85	70	316	286	898	5	56.8
95	80	324	294	892	5	56.5
105	90	332	302	886	4	56.3
115	100	338	309	881	4	56.0
140	125	353	325	868	3	55.5

The Absolute Pressure column from 18.7 to 140 is marked "INCHES OF MERCURY". The Gage Reading at Sea Level column from 4 to 125 is marked "VACUUM-INCHES OF MERCURY".

Table A.24. EQUIVALENT CAPACITIES OF PIPES OF SAME LENGTH—STEAM

NO. OF SMALL PIPES EQUIVALENT TO ONE LARGE PIPE

Size	1/2"	3/4"	1"	1 1/4"	1 1/2"	2"	2 1/2"	3"	3 1/2"	4"	5"	6"
1/2"	1.00	2.27	4.88	10.0	15.8	31.7	52.9	96.9	140	205	377	620
3/4"		1.00	2.05	4.30	6.97	14.0	23.3	42.5	65	90	166	273
1"			1.00	2.25	3.45	6.82	11.4	20.9	30	44	81	133
1 1/4"				1.00	1.50	3.10	5.25	9.10	12	19	37	68
1 1/2"					1.00	2.00	3.34	6.13	9	13	23	39
2"						1.00	1.67	3.06	4.5	6.5	11.9	19.6
2 1/2"							1.00	1.82	2.70	3.87	7.12	11.7
3"								1.00	1.50	2.12	3.89	6.39
3 1/2"									1.00	1.25	2.50	4.25
4"										1.00	1.84	3.02
5"											1.00	1.65
6"												1.00

continues

Table A.24. EQUIVALENT CAPACITIES OF PIPES OF SAME LENGTH—STEAM *(Cont'd)*

This table may be used to find the number of smaller pipes equivalent in steam-carrying capacity to one larger pipe. It may also be used to find the size of a larger pipe equivalent to several smaller ones. The pipes in either case must be of the same lengths.

EXAMPLE 1—Find the number of 1" pipes each 50 ft. long equivalent to one 4" pipe 50 ft. long.

SOLUTION 1—Follow down column headed 4" to the point opposite 1" in vertical column, and you will find that it will take 44 of the 1" pipes in parallel to equal one 4" pipe in steam-carrying capacity.

EXAMPLE 2—Find the size of one pipe equivalent to four 2" pipes in steam-carrying capacity.

SOLUTION 2—Find 2" in vertical column headed "Size" and follow across horizontally until closest number to 4 is found. The nearest to 4 is 4.5. Following this column up you will find that the size is 3 1/2". One 3 1/2" pipe is, therefore, equivalent in steam-carrying capacity to approximately four 2" pipes.

Table A.25. CAPACITY OF ROUND STORAGE TANKS

Depth or Length	NUMBER OF GALLONS									
	18	24	30	36	12	48	54	60	66	72
1 In.	1.10	1.96	3.06	4.41	5.99	7.83	9.91	12.24	14.81	17.62
2 Ft.	26	47	73	105	144	188	238	294	356	423
2 1/2	33	59	91	131	180	235	298	367	445	530
3	40	71	100	158	216	282	357	440	534	635
3 1/2	46	83	129	184	252	329	416	513	623	740
4	53	95	147	210	288	376	475	586	712	846
4 1/2	59	107	165	238	324	423	534	660	800	952
5	66	119	181	264	360	470	596	734	899	1057
5 1/2	73	130	201	290	396	517	655	808	979	1163
6	79	141	219	315	432	564	714	880	1066	1268
6 1/2	88	155	236	340	468	611	770	954	1156	1374
7	92	165	255	368	504	658	832	1028	1244	1480
7 1/2	99	179	278	396	540	705	889	1101	1335	1586
8	106	190	291	423	576	752	949	1175	1424	1691
9	119	212	330	476	648	846	1071	1322	1599	1903
10	132	236	366	529	720	940	1189	1463	1780	2114
12	157	282	440	634	684	1128	1428	1762	2133	2537
14	185	329	514	740	1008	1316	1666	2056	2490	2960
16	211	376	587	846	1152	1504	1904	2350	2844	3383
18	238	423	660	952	1296	1692	2140	2640	3200	3806
20	264	470	734	1057	1440	1880	2380	2932	3556	4230

RECTANGULAR TANK

To find the capacity in U.S. gallons of rectangular tanks, reduce all dimensions to inches, then multiply the length by the width by the height and divide the product by 231.

Example: Tank 56" long × 32" wide × 20" deep
Then 56" × 32" × 20" = 35840 cu. in.
35840 ÷ 231 = 155 gals. capacity

Table A.26. SYMBOLS FOR PIPE FITTINGS COMMONLY USED IN DRAFTING

Symbols courtesy of Mechanical Contractors Association of America, Inc.

	Flanged	Screwed	Bell and Spigot	Welded	Soldered
Bushing					
Cap					
Cross Reducing					
Straight Size					
Crossover					
Elbow 45-Degree					
90-Degree					
Turned Down					
Turned Up					
Base					
Double Branch					
Long Radius					
Reducing					
Side Outlet (Outlet Down)					
Side Outlet (Outlet Up)					
Street					

Table A.26. SYMBOLS FOR PIPE FITTINGS COMMONLY USED IN DRAFTING *(Cont'd.)*

	Flanged	Screwed	Bell and Spigot	Welded	Soldered
Joint Connecting Pipe					
Expansion					
Lateral					
Orifice Plate					
Reducing Flange					
Plugs Bull Plug					
Pipe Plug					
Reducer Concentric					
Eccentric					
Sleeve					
Tee Straight Size					
(Outlet Up)					
(Outlet Down)					
Double Sweep					
Reducing					
Single Sweep					
Side Outlet (Outlet Down)					
Side Outlet (Outlet Up)					
Union					

continues

Table A.26. SYMBOLS FOR PIPE FITTINGS COMMONLY USED IN DRAFTING *(Cont'd)*

	Flanged	Screwed	Bell and Spigot	Welded	Soldered
Angle Valve					
Check, also Angle Check					
Gate, also Angle Gate (Elevation)					
Gate, also Angle Gate (Plan)					
Globe, also Angle Globe (Elevation)					
Globe (Plan)					
Automatic Valve					
By-Pass					
Governor-Operated					
Reducing					
Check Valve (Straight Way)					
Cock					

Table A.26. SYMBOLS FOR PIPE FITTINGS COMMONLY USED IN DRAFTING *(Cont'd)*

	Flanged	Screwed	Bell and Spigot	Welded	Soldered
Diaphragm Valve					
Float Valve					
Gate Valve*					
Motor-Operated					
Globe Valve					
Motor-Operated					
Hose Valve, also Hose Globe					
Angle, also Hose Angle					
Gate					
Globe					
Lockshield Valve					
Quick Opening Valve					
Safety Valve					

* Also used for General Stop Valve Symbol when amplified by specification

Table A.27. TAP AND DRILL SIZES
(American Standard Coarse)

SIZE OF DRILL	SIZE OF TAP	THREADS PER INCH	SIZE OF DRILL	SIZE OF TAP	THREADS PER INCH
7	$1/4$	20	$49/64$	$7/8$	9
F	$5/16$	18	$53/64$	$15/16$	9
$5/16$	$3/8$	16	$7/8$	1	8
U	$7/16$	14	$63/64$	$1 1/8$	7
$27/64$	$1/2$	13	$1 7/64$	$1 1/4$	7
$31/64$	$9/16$	12	$1 13/64$	$1 3/8$	6
$17/32$	$5/8$	11	$1 11/32$	$1 1/2$	6
$19/32$	$11/16$	11	$1 29/64$	$1 5/8$	$5 1/2$
$21/32$	$3/4$	10	$1 9/16$	$1 3/4$	5
$23/32$	$13/16$	10	$1 11/16$	$1 7/8$	5
			$1 25/32$	2	$4 1/2$

Table A.28. WATER PRESSURE TO FEET HEAD

Pounds Per Square Inch	Feet Head	Pounds Per Square Inch	Feet Head
1	2.31	10	23.09
2	4.62	15	34.63
3	6.93	20	46.18
4	9.24	25	57.72
5	11.54	30	69.27
6	13.85	40	92.36
7	16.16	50	115.45
8	18.47	60	138.54
9	20.78	70	161.63

Table A.28. WATER PRESSURE TO FEET HEAD *(Cont'd)*

Pounds Per Square Inch	Feet Head	Pounds Per Square Inch	Feet Head
80	184.72	200	461.78
90	207.81	250	577.24
100	230.90	300	692.69
110	253.98	350	808.13
120	277.07	400	922.58
130	300.16	500	1154.48
140	323.25	600	1385.39
150	346.34	700	1616.30
160	369.43	800	1847.20
170	392.52	900	2078.10
180	415.61	1000	2309.00

NOTE: One pound of pressure per square inch of water equals 2.31 feet of water at 62° Fahrenheit. Therefore, to find the feet head of water for any pressure not given in the table above, multiply the pressure pounds per square inch by 2.31.

Table A.29. FEET HEAD OF WATER TO psi

Feet Head	Pounds Per Square Inch	Feet Head	Pounds Per Square Inch
1	.43	100	43.31
2	.87	110	47.64
3	1.30	120	51.97
4	1.73	130	56.30
5	2.17	140	60.63
6	2.60	150	64.96
7	3.03	160	69.29
8	3.46	170	73.63
9	3.90	180	77.96
10	4.33	200	86.62
15	6.50	250	108.27
20	8.66	300	129.93
25	10.83	350	151.58
30	12.99	400	173.24
40	17.32	500	216.55
50	21.65	600	259.85
60	25.99	700	303.16
70	30.32	800	346.47
80	34.65	900	389.78
90	38.98	1000	433.00

NOTE: One foot of water at 62° Fahrenheit equals .434 pounds pressure per square inch. To find the pressure per square inch for any feet head not given in the table above, multiply the feet head by .434.

Table A.30. PROPERTIES OF WATER AT VARIOUS TEMPERATURES

Temperature		Pressure of saturated vapor	Specific volume		Density Specific Weight		Conversion factor	Kinematic viscosity centistokes
°F	°C	lb/in² abs	ft³/lb	gal/lb	lb/ft³	g/cm³*	ft/lb/in²	
32	0	0.06859	0.016022	0.1199	62.414	0.9996	2.307	1.79
33	0.6	0.09223	0.016021	0.1196	62.418	0.9999	2.307	1.75
34	1.1	0.09600	0.016021	0.1196	62.418	0.9999	2.307	1.72
35	1.7	0.09991	0.016020	0.1196	62.420	0.9999	2.307	1.68
36	2.2	0.10395	0.016020	0.1196	62.420	0.9999	2.307	1.66
37	2.8	0.10615	0.016020	0.1196	62.420	0.9999	2.307	1.63
38	3.3	0.11249	0.016019	0.1196	62.425	1.0000	2.307	1.60
39	3.9	0.11696	0.016019	0.1196	62.425	1.0000	2.307	1.56
40	4.4	0.12163	0.016019	0.1196	62.425	1.0000	2.307	1.54
41	5	0.12645	0.016019	0.1196	62.426	1.0000	2.307	1.52
42	5.6	0.13143	0.016019	0.1196	62.426	1.0000	2.307	1.49
43	6.1	0.13659	0.016019	0.1196	62.426	1.0000	2.307	1.47
44	6.7	0.14192	0.016019	0.1196	62.426	1.0000	2.307	1.44
45	7.2	0.14744	0.016020	0.1196	62.42	0.9999	2.307	1.42
46	7.8	0.15314	0.016020	0.1196	62.42	0.9999	2.307	1.39
47	8.3	0.15904	0.016021	0.1196	62.42	0.9999	2.307	1.37
48	8.9	0.16514	0.016021	0.1196	62.42	0.9999	2.307	1.35
49	9.4	0.17144	0.016022	0.1196	62.41	0.9996	2.307	1.33
50	10	0.17796	0.016023	0.1199	62.41	0.9996	2.307	1.31
51	10.6	0.18469	0.016023	0.1199	62.41	0.9996	2.307	1.28
52	11.1	0.19165	0.016024	0.1199	62.41	0.9997	2.307	1.26
53	11.7	0.19883	0.016025	0.1199	62.40	0.9996	2.308	1.24
54	12.2	0.20625	0.016026	0.1199	62.40	0.9996	2.308	1.22
55	12.8	0.21392	0.016027	0.1199	62.39	0.9995	2.308	1.20
56	13.3	0.22183	0.016028	0.1199	62.39	0.9994	2.308	1.19
57	13.9	0.23000	0.016029	0.1199	62.39	0.9994	2.308	1.17

(*) Approximately numerically equal to specific gravity basis temperature reference of 39.2°F (4°C).

continues

Table A.30. PROPERTIES OF WATER AT VARIOUS TEMPERATURES (Cont'd)

Temperature		Pressure of saturated vapor	Specific volume		Density Specific Weight		Conversion factor	Kinematic viscosity centistokes
°F	°C	lb/in² abs	ft³/lb	gal/lb	lb/ft³	g/cm³ *	ft/lb/in²	
58	14.4	0.23843	0.016031	0.1199	62.38	0.9993	2.308	1.16
59	15	0.24713	0.016032	0.1199	62.38	0.9992	2.309	1.14
60	15.6	0.25611	0.016033	0.1199	62.37	0.9991	2.309	1.12
62	16.7	0.27494	0.016036	0.1200	62.36	0.9989	2.309	1.09
64	17.8	0.29497	0.016039	0.1200	62.35	0.9988	2.310	1.06
66	18.9	0.31626	0.016043	0.1200	62.33	0.9985	2.310	1.03
68	20	0.33889	0.016046	0.1200	62.32	0.9983	2.311	1.00
70	21.1	0.36292	0.016050	0.1201	62.31	0.9961	2.311	0.98
75	23.9	0.42964	0.016060	0.1201	62.27	0.9974	2.313	0.90
80	26.7	0.50683	0.016072	0.1202	62.22	0.9967	2.314	0.85
85	29.4	0.59583	0.016085	0.1203	62.17	0.9959	2.316	0.81
90	32.2	0.69613	0.016099	0.1204	62.12	0.9950	2.318	0.76
95	35	0.81534	0.016144	0.1205	62.06	0.9941	2.320	0.72
100	37.8	0.94924	0.016130	0.1207	62.00	0.9931	2.323	0.69
110	43.3	1.2750	0.016165	0.1209	61.96	0.9910	2.326	0.61
120	48.9	1.6927	0.016204	0.1212	61.71	0.9666	2.333	0.57
130	54.4	2.2230	0.016247	0.1215	61.56	0.9660	2.340	0.51
140	60	2.8892	0.016293	0.1219	61.38	0.9832	2.346	0.47
150	65.6	3.7184	0.016343	0.1223	61.19	0.9802	2.353	0.44
160	71.1	4.7414	0.016395	0.1226	60.99	0.9771	2.361	0.41
170	76.7	5.9926	0.016451	0.1231	60.79	0.9737	2.369	0.36
180	82.2	7.5110	0.016510	0.1235	60.57	0.9703	2.377	0.36
190	87.8	9.340	0.016572	0.1240	60.34	0.9666	2.3861	0.33
200	93.3	11.526	0.016637	0.1245	60.11	0.9628	2.396	0.31
210	98.9	14.123	0.016705	0.1250	59.86	0.9589	2.406	0.29
212	100.0	14.696	0.016719	0.1251	59.81	0.9580		

(*) Approximately numerically equal to specific gravity basis temperature reference of 39.2°F (4°C).

Table A.30. PROPERTIES OF WATER AT VARIOUS TEMPERATURES (Cont'd)

Temperature		Pressure of saturated vapor	Specific volume		Density Specific Weight		Conversion factor	Kinematic viscosity centistokes
°F	°C	lb/in² abs	ft³/lb	gal/lb	lb/ft³	g/cm³*	ft/lb/in²	
220	104.4	17.186	0.016775	0.1255	59.61	0.9549	2.416	
230	110.	20.779	0.016849	0.1260	59.35	0.9507	2.426	
240	115.6	24.968	0.016926	0.1266	59.08	0.9464	2.437	
250	121.1	29.825	0.017006	0.1272	58.80	0.9420	2.449	0.24
260	126.7	35.427	0.017089	0.1278	58.52	0.9374	2.461	
270	132.2	41.856	0.017175	0.1285	58.22	0.9327	2.473	
280	137.8	49.200	0.017264	0.1291	57.92	0.9279	2.486	
290	143.3	57.550	0.01736	0.1299	57.60	0.9228	2.500	
300	148.9	67.005	0.01745	0.1305	57.31	0.9180	2.513	0.20
310	154.4	77.667	0.01755	0.1313	56.98	0.9128	2.527	
320	160	89.643	0.01766	0.1321	56.63	0.9071	2.543	
330	165.6	103.045	0.01776	0.1329	56.31	0.9020	2.557	
340	171.1	117.992	0.01787	0.1337	55.96	0.8964	2.573	
350	176.7	134.604	0.01799	0.1346	55.59	0.8904	2.591	0.17
360	182.2	153.010	0.01811	0.1355	55.22	0.8845	2.608	
370	187.8	173.339	0.01823	0.1364	54.84	0.8787	2.625	
380	193.3	195.729	0.01836	0.1374	54.47	0.8725	2.644	
390	198.9	220.321	0.01850	0.1384	54.05	0.8659	2.664	
400	204.4	247.259	0.01864	0.1394	53.65	0.8594	2.684	0.15
410	392.2	276.694	0.01878	0.1404	53.25	0.8530	2.704	
420	215.6	306.780	0.01894	0.1417	52.80	0.8458	2.727	
430	221.1	343.674	0.01909	0.1426	52.38	0.8391	2.749	
440	226.7	381.54	0.01926	0.1441	51.92	0.8317	2.773	
450	232.2	422.55	0.01943	0.1453	51.47	0.8244	2.796	0.14
460	237.8	466.87	0.01961	0.1467	50.99	0.8169	2.824	
470	243.3	514.67	0.01960	0.1481	50.51	0.8090	2.851	

(*) Approximately numerically equal to specific gravity basis temperature reference of 39.2°F (4°C).

continues

Table A.30. PROPERTIES OF WATER AT VARIOUS TEMPERATURES *(Cont'd)*

Temperature		Pressure of saturated vapor	Specific volume		Density Specific Weight		Conversion factor	Kinematic viscosity centistokes
°F	°C	lb/in² abs	ft³/lb	gal/lb	lb/ft³	g/cm³*	ft/lb/in²	
480	248.9	566.15	0.02000	0.1496	50.00	0.8010	2.880	
490	254.4	621.48	0.02021	0.1512	49.48	0.7926	2.910	
500	260	680.86	0.02043	0.1528	48.95	0.7841	2.942	0.13
510	265.6	744.47	0.02067	0.1546	48.38	0.7750	2.976	
520	271.1	812.53	0.02091	0.1564	47.82	0.7661	3.011	
530	276.7	885.23	0.02118	0.1584	47.21	0.7563	3.050	
540	282.2	962.79	0.02146	0.1605	46.60	0.7485	3.090	
550	287.8	1045.43	0.02176	0.1628	45.96	0.7362	3.133	0.12
560	293.3	1133.38	0.02207	0.1651	45.31	0.7258	3.178	
570	298.9	1226.88	0.02242	0.1677	44.60	0.7145	3.228	
580	304.4	1326.17	0.02279	0.1705	43.86	0.7029	3.281	
590	310	1431.5	0.02319	0.1735	43.12	0.6906	3.339	
600	315.6	1543.2	0.02364	0.1768	42.30	0.6776	3.404	0.12
610	321.1	1661.6	0.02412	0.1804	41.46	0.6641	3.473	
620	326.6	1786.9	0.02466	0.1845	40.55	0.6496	3.551	
630	332.2	1919.5	0.02526	0.1890	39.59	0.6342	3.637	
640	337.8	2059.9	0.02595	0.1941	36.54	0.6173	3.737	
650	343.3	2203.4	0.02674	0.2000	37.40	0.5991	3.851	
670	354.4	2532.2	0.02884	0.2157	34.67	0.5554	4.153	
690	365.6	2895.7	0.03256	0.2436	30.71	0.4920	4.689	
700	371.1	3094.3	0.03662	0.2739	27.31	0.4374	5.273	
705.47	374.15	3206.2	0.05078	0.3799	19.69	0.3155	7.312	

(*) Approximately numerically equal to specific gravity basis temperature reference of 39.2°F (4°C).

Courtesy Schlumberger Industries, Inc.

Table A.31. RELATIONS OF ALTITUDE, PRESSURE, AND BOILING POINT

Altitude Feet	ATMOSPHERIC PRESSURE ABSOLUTE		BOILING POINT OF WATER °F (BASE PRESSURE PSI)				
	Inches of Mercury (Barometer)	Lbs. per Sq. In.	0	1	5	10	15
−500	30.46	14.96	212.8	216.1	227.7	239.9	250.2
−100	30.01	14.74	212.3	215.5	227.2	239.4	249.9
Sea Level	29.90	14.69	212.0	215.3	227.0	239.3	249.7
500	29.35	14.42	211.0	214.4	226.3	238.7	249.2
1000	28.82	14.16	210.1	213.5	225.5	238.1	248.6
1500	28.30	13.90	209.4	212.7	225.0	237.6	248.2
2000	27.78	13.65	208.2	211.7	224.1	236.8	247.7
2500	27.27	13.40	207.3	210.9	223.4	236.3	247.2
3000	26.77	13.15	206.4	210.1	222.7	235.7	246.7
3500	26.29	12.91	205.5	209.2	222.1	235.1	246.2
4000	25.81	12.68	204.7	208.4	221.4	234.6	245.7
4500	25.34	12.45	203.7	207.5	220.7	234.0	245.2
5000	24.88	12.22	202.7	206.8	220.1	233.4	244.7
6000	23.98	11.78	200.9	205.0	218.7	232.4	243.8
7000	23.11	11.35	199.1	203.3	217.3	231.3	242.9
8000	22.28	10.94	197.4	201.6	216.1	230.3	242.0
9000	21.47	10.55	195.7	200.0	214.8	229.3	241.3
10000	20.70	10.17	194.0	198.4	213.5	228.3	240.4
11000	19.95	9.80	192.2	196.8	212.3	227.3	239.6
12000	19.23	9.45	190.6	195.2	211.1	226.3	238.7
13000	18.53	9.10	188.7	193.6	209.9	225.4	237.9
14000	17.86	8.77	187.2	192.3	208.8	224.5	237.2
15000	17.22	8.46	185.4	190.6	207.6	223.6	236.4

Table A.32. WATER REQUIRED FOR HUMIDIFICATION

The approximate rule for calculating the amount of water required per hour to maintain any desired humidity in a room is: Multiply the difference between the number of grains of moisture per cubic foot of air at the required room temperature and humidity and the number of grains per cubic foot of outside air at the given temperature and humidity by the cubic contents of the room by the number of air changes per hour and divide the result by 7000 (this method disregards the expansion of air when heated.)

For most localities, it is customary to assume the average humidity of outside air as 30–40%.

Example:

Grains of moisture at 70° & 40% humidity =	3.19
Grains of moisture at 0° & 30% humidity =	0.17
Grains of moisture to be added per cu. ft. =	3.02

Assuming two air changes per hour in a room containing 8000 cu. ft. we have

$$\frac{(3.02 \times 8000 \times 2)}{7000} = 6.9 \text{ lbs. of water per hour required.}$$

Table A.33. WEIGHT OF WATER PER CUBIC FOOT AND HEAT UNITS IN WATER BETWEEN 32° AND 212°F.

Temperature Degrees F	Weight in pounds per Cubic foot	Heat Units (B.T.U.)	Temperature Degrees F	Weight in pounds per Cubic foot	Heat Units (B.T.U.)	Temperature Degrees F	Weight in pounds per Cubic foot	Heat Units (B.T.U.)
32	62.42	0.00	94	62.09	62.02	154	61.10	121.88
34	62.42	2.01	96	62.07	64.01	156	61.06	123.88
36	62.42	4.03	98	62.05	66.01	158	61.02	125.88
38	62.42	6.04	100	62.02	68.00	160	60.98	127.88
40	62.42	8.05	102	62.00	69.99	162	60.94	129.87
42	62.42	10.06	104	61.97	71.99	164	60.90	131.87
44	62.42	12.07	106	61.95	73.99	166	60.85	133.88
46	62.42	14.07	108	61.92	75.97	168	60.81	135.88
48	62.41	16.07	110	61.89	77.97	170	60.77	137.88
50	62.41	18.08	112	61.86	79.96	172	60.73	139.88
52	62.40	20.08	114	61.83	81.96	174	60.68	141.88
54	62.40	22.08	116	61.80	83.95	176	60.64	143.88
56	62.39	24.08	118	61.77	85.94	178	60.59	145.89
58	62.38	26.08	120	61.74	87.94	180	60.55	147.89
60	62.37	28.08	122	61.70	89.93	182	60.50	149.89
62	62.36	30.08	124	61.67	91.93	184	60.46	151.90
64	62.35	32.08	126	61.63	93.92	186	60.41	153.90
66	62.34	34.08	128	61.60	95.92	188	60.37	155.91
68	62.33	36.08	130	61.56	97.91	190	60.32	157.91
70	62.31	38.08	132	61.52	99.91	192	60.27	159.92
72	62.30	40.07	134	61.49	101.90	194	60.22	161.92
74	62.28	42.07	136	61.45	103.90	196	60.17	163.93
76	62.27	44.06	138	61.41	105.90	198	60.12	165.94
78	62.25	46.06	140	61.37	107.89	200	60.07	167.95
80	62.23	48.05	142	61.34	109.89	202	60.02	169.95
82	62.21	50.05	144	61.30	111.89	204	59.97	171.96
84	62.19	52.04	146	61.26	113.89	206	59.92	173.97
86	62.17	54.04	148	61.22	115.88	208	59.87	175.98
88	62.15	56.03	150	61.18	117.88	210	59.82	177.99
90	62.13	58.03	152	61.14	119.88	212	59.76	180.00
92	62.11	60.02						

Weights from Trans. A.S.M.E., Vol. 6

Heat Units from Goodenough's Steam Tables (1917).

INDEX